Tourism, Safety and COVID-19

This book offers empirical insights on key challenges faced by the travel and tourism industries in the post-COVID-19 era. The desire to make tourism safe is gaining ground, but what does this mean? This book explores the guarantees travelers want in the postpandemic era and how individual territories are predicting and responding to these needs. It explores the role of innovation and digital solutions, assures tourists different ways of using services, both physical and digital. It considers how the commitment of smart tourist cities to technology, sustainability and accessibility is able not only to improve the quality of travelers' tourist experience, but also the quality of life of local inhabitants. This book considers the main solutions that many destinations are already experimenting, around the world to respond to the new safety demands of travelers.

Salvatore Monaco is Postdoctoral Researcher in Sociology at the Faculty of Education, Free University of Bozen—Bolzano (Italy), where he teaches "Gender, Identities and Spaces." He obtained his PhD degree in Social Sciences and Statistics from the Department of Social Sciences, University of Naples Federico II (Italy). He is researcher at Osservatorio LGBT and OUT (Osservatorio Universitario sul Turismo), University of Naples Federico II. His research interests concern social exclusion, tourism, urban contexts, technologies and new media, with particular attention to issues related to identities, genders, sexual orientations and generations.

Routledge Insights in Tourism Series

This series provides a forum for cutting-edge insights into the latest developments in tourism research. It offers high-quality monographs and edited collections that develop tourism analysis at both theoretical and empirical levels.

Millennials, Spirituality and Tourism
Edited by Sandeep Kumar Walia and Aruditya Jasrotia

Tourism, Safety and COVID-19
Security, Digitization and Tourist Behaviour
Salvatore Monaco

Tourism, Safety and COVID-19

Security, Digitization and Tourist Behaviour

Salvatore Monaco

Routledge
Taylor & Francis Group
LONDON AND NEW YORK

First published 2022
by Routledge
2 Park Square, Milton Park, Abingdon, Oxon OX14 4RN

and by Routledge
605 Third Avenue, New York, NY 10158

Routledge is an imprint of the Taylor & Francis Group, an informa business

British Library Cataloguing-in-Publication Data
A catalogue record for this book is available from the British Library

Library of Congress Cataloging-in-Publication Data
A catalog record has been requested for this book

ISBN: 978-1-032-04920-5 (hbk)
ISBN: 978-1-032-04921-2 (pbk)
ISBN: 978-1-003-19517-7 (ebk)

DOI: 10.4324/9781003195177

Typeset in Times New Roman
by KnowledgeWorks Global Ltd.

Contents

Figures

Acknowledgments

I am grateful to Prof. Fabio Corbisiero, Prof. DeMond Miller and Prof. Urban Nothdurfter for their support and inspiring suggestions.

I would also like to acknowledge all my colleagues of OUT (Osservatorio Universitario sul Turismo) in the Departments of Social Science at University of Naples Federico II for their enthusiasm and encouragement during the writing of this book.

I am also indebted to Guido Guarino, Jo Malcolm, Marco Petrarca, Steven Ding and Titti Torre for their precious support.

Finally, I would like to thank the editors and reviewers at Routledge who have offered invaluable professional advice and assistance throughout the project.

Preface: Security, technology and travel in an age of crisis management

Travel amid the postdisaster landscape

The outbreak of the coronavirus beginning in 2019 will remain one of the most important and impactful social, political, economic events of the twenty-first century. *Tourism, Safety and COVID-19: Security, Digitization and Tourist Behaviour*, offers an insightful thought-provoking glimpse into future possibilities as we navigate through the COVID-19 pandemic and plan, prepare for, and mitigate against disasters yet to come. One of the more enduring characteristics of this crisis is the "emergence of social stress and anxiety due to the unfamiliarity and ambiguity of the phenomena, disruption in institutions' order and functional, and reduced social interactions" (Mansouri and Sefidgarbaei, 2021: 36). Unlike many of the disasters in the recent decades leading up to the COVID-19 pandemic—that is, hurricane Katrina (see Miller, 2008; Miller, Gonzalez and Hutter, 2017), Ebola outbreak, or Fukushima Daiichi nuclear disaster—a series of factors intersected to bring global mobility systems to a virtual standstill for "nonessential" travel. Even with the safety and stability of the transport sector called into question with the well-coordinated passenger airliner terror attacks at the beginning of the millennium with 09/11, in New York City, The Pentagon, and Shanksville, Pennsylvania, large sections of global travel remained open. Further threats to travel safety resulting from terror attacks in Madrid, Spain (train bombings on March 11, 2004[1]); and the 2016 Brussels Airport bombing[2] incidents, or natural hazards and climate change serve to disrupt travel across regions.[3]

In the 21st century, the boundaries between risk and fear are dissolving and risk, even perceived risks associated with daily life, are permeating the realm of travel. A fear of travel, arose as early as 2002, with the spread of severe acute respiratory syndrome (SARS). In 2014, the outbreak of the Ebola epidemic in West Africa triggered worldwide tourists' travel anxiety amid additional regional threats from Ebola. As noted by The World Travel & Tourism Council, "Although 99 percent of the Ebola cases were in Sierra Leone, Guinea and Liberia, many tourists regarded all African countries as high-risk destinations, resulting in a 7.7 percent decrease in arrivals in other Ebola-free countries" (World Travel & Tourism Council, 2018; Zheng,

Luo and Ritchie, 2021: 1). The fear of contracting Ebola via contact with travelers impacted travel in both the United States and Europe in 2014 until the contemporary COVID-19 crisis. The initial outbreak and lockdowns, followed by subsequent waves and additional global, regional and local travel restrictions, severely curtailed travel, tourism and tourism-related activities. Moreover, due to the uncertainty, fear, and misinformation of the pandemic itself and travel to physical locations decreased markedly. In the postnormal period of the COVID pandemic disaster, the initial shock of navigating the new disaster landscape and negotiating the risks to personal security in an age of mobility instability triggered by incalculable and unpredictable public health challenges that significantly impact tourists' choice to travel as well as destinations and activities.

New risks, new modalities and new opportunities

Tourism, Safety and COVID-19: Security, Digitization and Tourist Behaviour is an exciting book because it is written during what the theorist Ziauddin Sadar calls postnormal times. In Sadar's work, he proclaims, "Welcome to postnormal times." He continues by further positing,

"[i]t's a time when little out there can be trusted or gives us confidence. The *espiritu del tiempo*, the spirit of our age, is characterized by uncertainty, rapid change, realignment of power, upheaval, and chaotic behavior. We live in an in-between period where old orthodoxies are dying, new ones have yet to be born, and very few things seem to make sense" (Sadar, 2009: 435).

While in some situations, "the very structure of society [is] temporarily suspended" (Szakolczai, 2009: 142), the notion of liminality may appear to be a state of stasis, it is anything but. During the liminal state, there is a sequence of rites and processes that help structure the period of postnormality. van Gennep's (1977) trajectory toward a new beginning starts where there is a recognition that the old way or the old orthodoxies cannot explain the current problem. In so many ways, the very act of travel is a liminal activity. COVID-19 has forced us to reflect on how to travel as a state of transitioning where participants transit from one space, place, or social status to another. In many ways, to travel is to be "betwixt and between" as being neither here nor there; they are betwixt and between the positions and at the same time being both (Turner, 1967). Thus, for van Gennep (1977) the act of separation of the old orthodoxies of movement between spaces, the metaphorical death or separation for the old life preliminal rites serves as a formal recognition of separation (rites of separation) from the previous understanding. Redefined mobility, in the era of COVID-19, came as an abrupt series of border closures that disrupted the flow. Soon, the second stage, liminal rites mark the transition with rules of travel and strict government measures and public guidance served to structure the middle existence of travel, movement, and tourism where the physical, mental and psychological change from the old orthodoxies and the new normal begins to take place. This second component of transition rites "implies an actual passing through the threshold that marks the boundary between two phases, and the term 'liminality' was introduced in order to characterize this passage" (Szakolczai, 2009: 141). Finally, the ability to incorporate the new understanding of what it means to be normal and the ability to begin again refers to postliminal rites where we are reincorporated into society and advancing into the new normal.

"According to Turner, all liminality must eventually dissolve, for it is a state of great intensity that cannot exist very long without some sort of structure to stabilize it... either the individual returns to the surrounding social structure... or else liminal communities develop their own internal social structure, a condition Turner calls *normative communitas*" (Homans, 1979: 207).

For Beck, the development of a society characterized by risk, fear, and ambiguity, or a risk society, where fear is the prevailing subjectivity, linked via uncertainty to what he called the "horror of ambiguity" (Beck, 2009: 5; see Ward, 2020). Ward lists several themes, in part here, that also characterize this period in between the old orthodoxies and the emergence of the new normal. Some of the themes Ward highlights are risks (Beck, 1992; Douglas, 1992; Beck, 2009; Luhmann, 2017), fear (Slovic, Fischhoff and Lichtenstein, 1990; Bauman, 2006), and crisis (Habermas, 1975; Berlant, 2011; Walby, 2015). I add two additional themes: exaggerated risks (Whipple, 1986) and chaos (Sardar, 1999). As a way to cope with the new fears that abide with the global pandemic, sheltering in place, social distancing, quarantine, mask-wearing, emergencies, first waves, second waves, new strains, variants, lockdowns, shutdowns, contact tracing, social distancing and hand washing all became common responses as behavior changes impacting how we interact with each other and the physical world have become normalized. The risks expressed as a result of contact with unseen pathogens, whether contact with humans or objects, could render illness or death and will have long-term effects on many aspects of the human interaction experience. Whether in public and private settings, at work and leisure, the "forced lockdowns have resulted in a grand experiment in 'work from home' and widespread use of videoconferencing and telecommuting that may result in a major shift in business travel behavior. Employers have already concluded that working from home brings them benefits as they need less offices and can staff outside major urban centers, translating into lower overheads" (Robinson, 2020: 7).

Traditional travel, and other aspects of tourism, requires humans to come into close contact, share intimate spaces with a stranger or multiple strangers, remain in long lines, and often confronting new environments. Such experiences with close unprotected contact in foreign places or places with limited medical facilities can pose a source of perceivable fear and security risk for travelers. Hence, the restrictions on mobility and perceived risks of travel during the COVID-19 pandemic has prevented travelers from the experience of travel bringing forth new forms of tourism experiences to connect tourists with destinations that transcend the liminal space between the old orthodoxies of tourism and the new and emergency forms of touristic experience without changing locations. Technologically enhanced tourism with customizable virtual components allows for "travel" to multiple locations without leaving the comforts and security of one's home. This form of technology infuses several themes (e.g., risk, fear, uncertainty) in prior research consider essential parts of the tourism experience and serves as a way to virtually have the experiential nature of tourism (Inkson and Minnaert, 2012) while mitigating against such issues. For example, the virtual experience and perception of that experience, all within the safety and security of one's home, opening a new era for the tourist-albeit mediated virtually by technology. Technology offers a new set of opportunities that will shape crisis, mobility and security to come.

Mobility, security and crisis in the disaster landscape

Safety and security remain necessary conditions for travel and tourism. In chapter three, "How to choose safe destinations: From 'communitycation' to new digital analysis tools," and chapter four, "A system in crisis: Means of transport in search of solutions and new functions to withstand the pandemic," *Tourism, Safety and COVID-19: Security, Digitization and Tourist Behaviour* sets out to address how crisis management in the era of COVID-19 will continue to shape our conception of safety and security by expanding beyond political security, public safety, knowable health and sanitation issues, personal data safety, legal protection of tourists, consumer protection, safety in communication, disaster protection, environmental security, to a more complex multidimensional (Kovari and Zimanyi, 2011) nuanced understanding of risks that are incorporated into crisis management strategies. The number of international tourists dropped by 98 percent in May 2020 when compared to the year before (UNWTO, 2020a) with a large number of those travelers not likely to return to the pre-COVID-19 travel flows (Lew *et al.*, 2020; Joo *et al.*, 2021). The traveler's emotional journey, including risk and perceived risks,[4] is inexorably linked to the physical or process journey and as the surge of COVID-19 dissipates, hence, taking into account the customer experience as in the reengineering of the entire travel experience continuum is important (Robinson, 2020). While scholars recognize that risk is inherent in decision making (Dowling and Staelin, 1994), risk has greater implications in the tourism industry (Quintal, Lee and Soutar, 2010; Yang and Nair, 2014); tourism managers must be able to understand how "...perceived risk is instrumental in understanding tourists' attitudes and behaviors. [Because] [f]or some tourists, a manageable degree of perceived risk can make their tourism experiences more stimulating and memorable" (Quintal, Lee and Soutar, 2010; Yang and Nair, 2014; Karl *et al.*, 2020). Robinson further contends that involving multiple interfaces and touchpoints in the journey from origin to destination. Securing the travel experience, end-to-end, based on digital passenger-innovation can overcome the fragmented travel ecosystem so that the traveling public can have a real-time data-driven experience with the traveler controlling some aspects of their risks, crisis and crisis management will require a new way of thinking about the travel experience and their data (Robinson, 2020).

Key in this understanding and more nuanced view of risk, due to both its tangible and intangible natures, is the ability to tangibilize the intangible aspects of risk for travelers in a post-COVID-19 world in much the same way we seek to tangibilize the scenic view, or the majesty of a mountain as part of the tourist package to build a certain level shared understanding so that the ability to manage such risks are not experienced but rather mitigated. In this way, security, namely public health security, remains a necessary component of travel while risks are linked to tangible safety that is knowable. Reddy, Buskirk and Kaicker (1993) notes that this is a complicated process that involves socially constructed understandings that are overemphasized by concrete elements, whereas an overemphasis on the abstract intangible elements, further compounds the intangibility. Crisis management allows for the integration of the tangible and intangible aspects of complex social concepts that ultimately drive tourist decision-making and behavior in the liminal transitory stage (postliminal rites) into a new normal for travel in the post-COVID-19 world.

New modalities to remain safe and connected

In so many ways, the goal of safety as both a tangible and intangible aspect of tourism has come to the forefront with primary examples that link technology with the existing physical spaces to create virtual places where patrons to museums or cultural events can remain engaged with the art or remain connected with the tour or remain connected with the location via virtual reality (VR) experience. As argued in chapter five "Being a tourist without moving: Stationary tourism as an alternative strategy for traveling," the use of VR as a primary, not augmentative approach to tourism in the wake of the pandemic and despite decades of clear objects by event managers and destination marketers, represents a clear mark in the realization and transformation of museum, place, and cultural heritage tourism. The use of such VR strategy, in hopes to retain traveler interest, employed by entire destinations are Egypt and Singapore destination campaign created by the Singapore Tourism Board (2020; El-Said and Aziz, 2021) designed to allow virtual visitors to explore some of the city-states as they... "Experience Singapore now. Visit later" as a way to remain in touch, as a way to help Singapore remain, as both a virtual and physical location to visit. Egypt's approach integrated new technology with social media platforms by launching the "Explore Egypt from Home" with VR tours launched by the Egyptian Ministry of Tourism and Antiquities that explore the tomb of Meresankh III, the tomb of Menna, the Ben Ezra Synagogue, the Red Monastery and the Mosque-Madrassa of Sultan Barquq (Machemer, 2020).

The trend to include VR is one of the first approaches in a series of steps needed to manage the crisis within the industry and the first reaction to keep the public engaged at a time when government authorities closed public spaces, borders, and events to nonessential travel to limit the spread of the virus. While the essence of nonessential and essential are clearly defined in the realm of public health, these roles become blurred within the scope of regional economic social, and economic wellbeing of regional and international tourist destinations. As such, Gössling, Scott and Hall (2020; Sigala, 2020; El-Said and Aziz, 2021) argue, the need for major long-term structural and transformational changes in the tourism eco-system following the COVID-19 pandemic and reform the travel and tourism sector so that it is more dependent on the use of technology (e.g., robotic applications, mobility tracking technologies, virtual reality applications, virtual tours, digital identity controls in airports and self-service check-in kiosks) as a safe alternative to direct human interactions (Sigala, 2020; Zeng, Chen and Lew, 2020), and to develop innovative and digitalized tourism experiences (UNWTO, 2020b).

In this way, travel and tourism can be more sustainable (Ioannides and Gyimóthy, 2020; UNWTO, 2020b; El-Said and Aziz, 2021) and viable for generations to come and be able to manage the crisis and crises of the future. However, other innovations in the sector will be needed to provide not only an authentic experience, but to also promote trust in the industry to commodify the tangible and intangible aspects of security that help not only the travelers, and the industry, but also the communities who rely on tourism by providing virtual experiences. For instance, virtual tour of the Faroe Islands (https://www.remote-tourism.com), which concluded July 2020, allowed virtual visitors some interactive control of their tour guide using a free app and have two minutes of control over the guide, who also provides a commentary of the experience. In this approach, camera-wearing locals will respond to sight-seeing commands from people at home, allowing virtual tourists to control their own route (Springwise, 2020a). Furthermore, contactless service, as public venues open, are now almost essential. With hospitality predicated on personal touches and unique experiences, the hospitality industry is also commodifying health security in innovative ways by introducing the contactless stay and other personal touches to allow the traveling public to feel more secure, through minimal or no contact, yet retain the key components of connectivity, individuality and safety in the postpandemic new normal (Springwise, 2020b). The "free to download" citizenM Hotel app offers a worldwide portfolio of locations. Furthermore, the app allows for room selection, check-in/check-out, entertainment, lighting controls, and optional room cleaning service (Springwise, 2020b). While other postpandemic innovations in air travel include reverse middle seating and more double-decker cabins (McHahon, 2020), seating re-designs that provide for a more individualized flight, or airplane seats that change color when sanitized (Stieg, 2020) offer ways to tangibilize security during flight. With many more innovative ways to enhance mobility that is safe to engage both physically, in the wake of disasters, crises and future pandemics, and virtually, where the travelers' personal, financial and medical data are equally secure.

When will we know if it's safe to travel again? This remains the question for many business and leisure travelers now planning to resume their mobile activities after the pandemic. With overall business and leisure traveler confidence reaching new lows during the pandemic, the future of travel remains clouded with perceived risks and uncertainty. As vaccines and vaccine passports become readily available, we will strike new opportunities to manage the current crisis and develop crisis management plans to restore the trust among members of the traveling public for future disasters.

Tourism, Safety and COVID-19: Security, Digitization and Tourist Behaviour is written at a time of great transformation in the tourism and hospitality industries and mobility, I encourage you to think about many of the larger issues posed as we embark on a new normal for travel. This book provides insights into topics of disaster and hazard management that emphasize the evolution of technology as the digitization of information becomes critical to a resilient travel sector. Each chapter focuses on issues that influence the ongoing development of mobility dynamics and factors across different communities, cultures, nations, and international relations. I challenge the readers to reflect upon the examples provided throughout this book so that we as a global community may move away from a responsive culture of emergency and crisis management practices and policies and toward mitigation strategies and policies that reduce disaster and hazard risk for the travel experience continuum.

DeMond S Miiller
Professor of Sociology
Rowan University (Glassboro, New Jersey)

Notes

1. 2004 Madrid train bombings against the Cercanías commuter train system three days before Spain's general elections with 191 fatalities and over 1,800 injured (Reinares, 2016).
2. On March 22, 2016 the Brussels Airport (at the commune of Zaventem) and Maelbeek Subway Station, located near the European District resulting in 35 fatalities (including the three attackers) and 300 injured.
3. For example, The Indian Ocean Tsunami, caused by a major earthquake on December 24, 2004 curtailed the entire tourism economy in Southeast Asia (impacting nations as far as Thailand, Maldives, Sri Lanka, Indonesia and

India). For instance, "The Egyptian tourism industry had been on the rise for nearly three decades before the Arab Spring... According to the Central Bank of Egypt (CBE) monthly tourist arrivals averaged 471,460 from 1982 until 2016 reaching an all-time high of 1.486 million in October 2010 and a record low of 57,000 in February 1991" (Tomazos, 2017); however, following the civil unrest and citizen uprisings in North Africa, particularly in Egypt, amid the Egyptian Arab Spring in 2011, tourist arrivals declined by nearly a third when compared to 2010, a record year for tourist arrivals in Egypt (Tomazos, 2017).

Bibliography

Bauman, Z (2000) *Liquid Modernity*, Cambridge: Polity Press.

Bauman, Z (2001) *The Individualized Society*, Cambridge: Polity Press.

Bauman, Z (2006) *Liquid Fear*, Cambridge: Polity Press.

Beck, U (1992) *Risk Society: Towards a New Modernity*, London: Sage.

Beck, U (2009) *World at Risk*, Cambridge: Polity Press.

Berlant, L (2011) *Cruel Optimism*, Durham: Duke University Press.

Douglas, M (1992) *Risk and Blame: Essays in Cultural Theory*, London: Routledge.

Dowling, G R and Staelin, R (1994) 'A model of perceived risk and intended risk-handling activity', *Journal of Consumer Research*, 21, 1: 119–134.

El-Said, O and Aziz, H (2021) 'Virtual Tours a means to an end: An analysis of virtual tours' role in tourism recovery post COVID-19', *Journal of Travel Research*, 1: 1–21.

Gössling, S, Scott, D and Hall, C (2020) 'Pandemics, tourism and global change: A rapid assessment of COVID-19', *Journal of Sustainable Tourism*, 1: 1–20.

Habermas, J (1975) *Legitimation Crisis*, Boston: Beacon Press.

Homans, P (1979) *Jung in Context: Modernity and the Making of a Psychology*, Chicago: University of Chicago Press.

Inkson, C and Minnaert, L (2012) *Tourism Management: An Introduction*, Thousand Oaks: Sage Publications.

Ioannides, D and Gyimóthy, S (2020) 'The COVID-19 crisis as an opportunity for escaping the unsustainable global tourism path', *Tourism Geographies*, 22, 3: 624–632.

Joo, D, Xu, W, Lee, J, Lee, C and Woosnam, K M (2021) 'Residents' perceived risk, emotional solidarity, and support for tourism amidst the COVID-19 pandemic', *Journal of Destination Marketing & Management*, 19: 100553.

Karl, M, Muskat, B and Ritchie, B W (2020) 'Which travel risks are more salient for destination choice? An examination of the tourist's decision-making process', *Journal of Destination Marketing & Management*, 18: 100487.

Kovari, I and Zimanyi, K (2011) 'Safety and security in the age of global tourism (the changing role and conception of safety and security in tourism)', *Applied Studies in Agribusiness and Commerce*, 5, 3: 59–61.

Lew, A, Cheer, J M, Haywood, M, Brouder, P and Salazar, N B (2020) 'Visions of travel and tourism after the global COVID-19 transformation of 2020', *Tourism Geographies*, 22, 3: 455–466.

Luhmann, N (2017) *Risk: A Sociological Theory*, New York: Routledge.

McHahon, S (2020) 'Plane cabins could change dramatically because of the pandemic. Here's how'. Retrieved from *Washington Post*: https://www.washingtonpost.com/travel/2020/08/18/plane-cabins-could-change-dramatically-because-pandemic-heres-how (accessed 04/15/2021).

Machemer, T (2020) 'Take a Free Virtual Tour of Five Egyptian Heritage Sites'. Retrieved from *Smithsonianmag.com*: https://www.smithsonianmag.com/smart-news/virtually-tour-five-egyptian-landmarks-180974696 (accessed 04/15/2021).

Mansouri, F and Sefidgarbaei, F (2021) 'Risk society and COVID-19', *Canadian Journal of Public Health*, 112: 36–37.

Miller, D S (2008) 'Disaster tourism and disaster landscape attractions after hurricane Katrina: An auto-ethnographic journey', *International Journal of Culture, Tourism and Hospitality Research*, 2, 2: 115–131.

Miller, D S, Gonzalez, C and Hutter, M (2017) 'Towards an understanding of Phoenix tourism within dark tourism: The rebirth, rebuilding and rebranding of tourist destinations in landscapes of risk and uncertainty following disasters', *Worldwide Hospitality and Tourism Themes*, 9, 3: 196–215.

Quintal, V A, Lee, J A and Soutar, G N (2010) 'Tourists' information search: The differential impact of risk and uncertainty avoidance', *International Journal of Tourism Research*, 12, 4: 321–333.

Reddy, A C, Buskirk, B D and Kaicker, A (1993) 'Tangibilizing the intangibles: Some strategies for services', *The Journal of Services Marketing*, 7, 3: 13.

Reinares, F (2016) *Al-Qaeda's Revenge: The 2004 Madrid Train Bombings*, New York: Columbia University Press.

Robinson, J (2020) 'Thoughts on the post-pandemic new normal in air travel', *Journal of Airport Management*, 15, 1: 6–19.

Sardar, Z (2010) 'Welcome to postnormal times', *Futures : The Journal of Policy, Planning and Futures Studies*, 42, 5: 435–444.

Sardar, Z (1999) *Introducing Chaos*, London: Icon Books.

Sigala, M (2020) 'Tourism and COVID-19: Impacts and implications for advancing and resetting industry and research', *Journal of Business Research*, 117: 312–321.

Singapore Tourism Board (2020) 'Virtual Experiences: Immerse in 360° Experiences from Your Home'. Retrieved from: https://www.visitsingapore.com/virtual-experiences/virtual-experiences-listing (accessed 04/15/2021).

Slovic, P, Fischhoff, B and Lichtenstein, S (1990) 'Facts versus fears: Understanding perceived risk', in D Kahnemann (ed.) *Judgement Under Uncertainty: Heuristics and Biases*, Cambridge: Cambridge University Press.

Springwise (2020a) 'A virtual tour of the Faroe Islands with remote-controlled guides'. Retrieved from: https://www.springwise.com/innovation/travel-tourism/remote-control-people-faroe-islands (accessed 04/15/2021).

Springwise (2020b) 'Hotel introduces Contactless stays via free app'. Retrieved from: https://www.springwise.com/innovation/travel-tourism/citizenm-app-contactless-stays-coronavirus (accessed 04/15/2021).

Stieg, C (2020) 'From seats that change color when clean to staggered rows, here's what airplane cabins could look like post-pandemic'. Retrieved from *CNBC Makeit*: https://www.cnbc.com/2020/08/08/photos-design-studio-re-imagined-plane-with-covid-19-safety-features.html (accessed 04/15/2021).

Szakolczai, A (2009) 'Liminality and experience: Structuring transitory situations and transformative events', *International Political Anthropology*, 2, 1: 141–172.

Tomazos, K (2017) 'Egypt's tourism industry and the Arab spring', in R Butler and W Suntikul (eds.) *Tourism and Political Change* (2nd ed.), Oxford: Goodfellow.

Turner, V (1967) *The Forest of Symbols*, Ithaca: Cornell University Press.

UNWTO (2020a) 'UNWTO World Tourism Barometer May 2020. Special Focus on the Impact of COVID-19'. Retrieved from: https://webunwto.s3.eu-west-1.amazonaws.com/s3fs-public/2020-05/Barometer%20-%20May%202020%20-%20Short.pdf (accessed 04/15/2021).

UNWTO (2020b) 'Impact of COVID-19 on global tourism made clear as UNWTO counts the cost of standstill. UNWTO'. Retrieved from: https://www.unwto.org/news/impact-of-covid-19-on-global-tourism-made-clear-as-unwto-counts-the-cost-of-standstill (accessed 04/15/2021).

van Gennep, A (1977) *The Rites of Passage*, London: Routledge and Kegan Paul.

Walby, S (2015) *Crisis*, Cambridge: Polity Press.

Ward, P R (2020) 'A sociology of the covid-19 pandemic: A commentary and research agenda for sociologists', *Journal of Sociology*, 56, 4: 726–735.

Whipple, C G (1986) 'Dealing with uncertainty about risk in risk management', in R W Kates and A M Weinberg (eds.) *Hazards: Technology and Fairness*, Washington: The National Academies Press.

World Travel & Tourism Council (2018) 'Impact of the Ebola Epidemic on Travel and Tourism'. Retrieved from: https://www.wttc.org/-/media/files/reports/2018/impact-of-the-ebola-epidemic-on-travel-and-tourism.pdf (accessed 04/15/2021).

Yang, C L and Nair, V (2014) 'Risk perception study in tourism: Are we really measuring perceived risk', *Procedia-Social and Behavioral Sciences*, 144, 1: 322–327.

Zeng, Z, Chen, P and Lew, A (2020) 'From high-touch to high-tech: COVID-19 drives robotics adoption', *Tourism Geographies*, 22, 3: 724–734.

Zheng, D, Luo, Q and Ritchie, B W (2021) 'The role of trust in mitigating perceived threat, fear, and travel avoidance after a pandemic outbreak: A multigroup analysis', *Journal of Travel Research*, 2: 1–20.

Introduction

Tourism and COVID-19, among desire to travel, new fears and opportunities

The aim of this book is to discuss the issue of mobility in contemporary society, with the specific objective of understanding the new features of tourism against the backdrop of the pandemic emergency. More specifically, the purpose of this book is to achieve an empirical understanding of the key challenges for the travel and tourism sectors in the post-COVID-19 era, focusing in particular on the relationship among tourism, security and digitization.

From a sociological point of view, studying social reality starting from tourism could be a useful investigation strategy. Tourism, in fact, is by its nature an interesting object of analysis that allows us to trace the changes that cross the social, economic, cultural and technological spheres. When we talk about tourism, we are referring to a phenomenon that is not only widespread globally by now, but which is also capable of gaining different forms and meanings over time, adapting to historical contingencies, changing its characteristics as a whole or in part. In other words, tourism evolves and changes as society does, being strongly influenced by the changes that the latter goes through. Thus, quoting Urry (1990: 25) it can be argued that "In relationship to tourism it is crucial to recognize how the consumption of tourist services is social [...]. Explaining the consumption of tourist services cannot be separated off from the social relations within which they are embedded."

In this regard, Urry and Sheller (2004) in the field of social research have proposed abandoning the traditional reading of social phenomena linked to a sedentary lifestyle and territoriality, by raising mobility as an interpretative key of the main characteristics of society. The different mobilities "are materially transforming social as society into social as mobility" (Urry, 2000: 186).

In particular, tourism mobility, far from being the kind of marginal, exclusive or peripheral activity it once was, becomes a massively popular activity in the contemporary life of social actors, shortening distances due to intersubjective interactions.

Among other things, the changes that have occurred in society have extended the tourist practice from the mere physical movement of people in relatively remote destinations into something much more articulated and complex.

DOI: 10.4324/9781003195177-1

Thus, in contemporary society it is actually possible to identify the new forms of travel that are taking place, along with the transfer of not only tourists, but also data, brands and information. Therefore, today social media-based tourism is made possible by new communication technologies, contributing significantly to experience-oriented alternative forms of tourism. The mobility achieved through new digital media represents a new principal element among the factors that extend the meaning of tourism activity. Digital mobilities allow tourists to be on the road while staying at home, but also staying at home while they are physically in another place (e.g., Urry, 2002; White and White, 2007). In this regard, the image of the "independent traveler" (Mascheroni, 2006; 2007) has been extended to the world of literature, in reference to a visitor around the world, but in constant contact with his place of origin.

Precisely for the aforementioned reasons, today the analysis of tourism can be a useful tool for an extended interpretation of social reality. In other words, the study of tourist mobility allows us to go beyond tourism itself, enabling us to categorize practices of daily living with the same logics that characterize mobility systems. In this sense, we can argue that the multiplication of tourist mobility has allowed tourism to embody the typical processes of society, a sort of mirror through which it is possible simultaneously to observe ourselves and even the transformations of contemporary society (Mancinelli and Palou, 2016).

Thus, the boundless, constant, continuous, incessant flows that characterize contemporary society represent the main noteworthy element for a full understanding of social reality; conversely, interpreting the phenomena circumscribing them to the territories within which they take place appears no longer adequate (Urry and Sheller, 2006). Taking this perspective represents a spatial turning point in the field of social research (Urry, 2003), since scholars can investigate individual and collective experiences with the awareness that they are the result of flows taking place on a global level (Rojek and Urry, 1997).

Furthermore, today, more than ever before, even immobility must be taken into account in the analysis of social reality, as it is complementary to mobility (Franquesa, 2011; Adey *et al.*, 2021) since we are living in an unprecedented time of change, in which the occurrence of COVID-19 represents a real watershed moment in history.

Although change has always been part of the human experience, the transformation sweeping across the globe in this recent period has been deteriorating and even breaking down the mechanisms of predictability. In fact, on a global level, the idea of a world in which science could help us to remove all obstacles, face any danger or problem and get through periods of crisis, was rooted in the collective imagination. The pandemic, however, has made it clear that science is actually not a panacea. The emergence and spread of COVID-19 were so rapid and sudden that the experts were unable to guarantee reassurance, showing their vulnerability in explaining the overall picture of the situation and the underlying contradictions of the problem. Doctors, scientists and experts around the world have revealed that science takes time to analyze sudden or unexpected phenomena as in the case of COVID-19. In this specific situation, unprecedented efforts have been made to find treatments and vaccines capable of dealing with the health problem in a timely manner.

This once-in-a-century global pandemic has left everyone shocked: not knowing what awaits us in the future has made us feel panic-stricken, we are forced to radically change most of our habits in order to avoid infection (e.g., Yu *et al.*, 2021). As a result, we have had to learn how to cope with a new daily routine, which is replacing our previous ones.

The pandemic that has hit the world is a phenomenon which is probably inexplicable through the theory of the black swan (Taleb, 2007), contrary to what some scholars claim (e.g., Farinella and Simula, 2020; Lüscher, 2020; Yarovaya, Matkovskyy and Jalan, 2020; Mazzoleni, Turchetti and Ambrosino, 2020; Morales and Andreosso-O'Callaghan, 2020; Antipova, 2021; Simianer and Reimer, 2021). This theory was formulated with reference to all those surprising events, which, once passed, are rationalized *a posteriori*, suggesting that they were predictable or explainable.

Although several studies in the past documented a long history of zoonoses that turned into pandemics (e.g., Jones *et al.*, 2008; Morse *et al.*, 2012; Han, Kramer and Drake, 2016; Schmeller, Courchamp and Killeen, 2020), I believe that COVID-19 and the disruptive effects it has had on all aspects of life, including tourism, were not so predictable.

In addition, there was a distorted vision of the reality beyond the pandemic event, precisely because of the presumption that the capitalist system can reduce nature and the biosphere into a technosphere, a sort of artificial product owned by man that was always measurable, predictable and manageable.

On the contrary, the world was caught so much by surprise that, the upheaval caused by COVID-19 has exerted and will exert significant effects also on our way of living in territories and moving around the world.

From this critical angle, it is safe to argue that the relationship between tourism and the pandemic is endowed with a double interpretation, which will be explored in this book in parallel. First, this relationship is a potential danger because the spread of the virus, in addition to being the cause of illness and death, showed its disruptive impact on the world economy in general and also on the tourism industry in particular, placing the sector in front of a series of unprecedented challenges. All the players in the sector, as a result, from travel agencies to accommodation facilities, transport companies to catering and entertainment, experienced an economic collapse, which was partly combatted also with timely innovative solutions and countermeasures. In this scenario, new technologies have played a central role since they have been surprisingly useful in confronting some of the main critical problems that the pandemic has posed.

The second interpretation explored in this book is a recognition that the tourism industry is not just a victim of the pandemic. Because of long-haul travel, mobility has been actually part of the problem itself. Without entering into specific medical-scientific discussions, we can argue that the deterioration of the pandemic crisis has been fueled by global competition among tourist destinations, the proliferation of information available to travelers, the better living and economic conditions for people and overall the increase in international travel (Knobler, 2006). In the contemporary world, travel is much more widespread than in the past, driven by differentiated motivations and directed toward a multiplicity of destinations, which today appear to be more and more easily accessible (Colleoni, 2010). This diversification fits within the ranks of postmodernity, in which social narratives are more individual and reflective, encouraged by the new means of transport and communication, as well as by the increase in possibility of choosing and purchasing opportunities. Moreover, these elements have made even more evident the extent to which in postmodern society people develop their paths of life and choose what they really want to be also through leisure and tourism (e.g., Rojek, 1997; Kaplan, 2000; Ritzer and Stillman, 2001; Haworth and Veal, 2004; Timothy and Olsen, 2006). Starting from these considerations, it is evident that the progressively increasing ease and speed of transfer of people and goods has undoubtedly contributed to the spread of the infection. As Iaquinto (2020: 175) suggests, it is possible to consider tourists as "vectors," since the presence of tourists in confined spaces such as that of cruise ships, airplanes, trains, tour buses or hostels, together with the tourists' practices in which close and prolonged physical contact is required, all facilitated "the transmission of COVID-19 and its elevation to a pandemic." In other words, the coronavirus crisis has made tangible how much the expansion of the global tourism market has undeniably increased the danger of exposure to new diseases, globally amplifying the risk of transmission (Baker, 2015).

Citizens around the world have realized the responsibilities of mobility and, for this reason, a sort of aversion toward mobility has gained ground, especially with regards to international travel. Recent studies (e.g., Zheng, Luo and Ritchie, 2021) have shown that the fear generated after the outbreak of the pandemic has evoked in people different coping strategies and the adoption of more cautious travel behaviors. We can argue, however, that people have not lost the desire to travel. In the first summer of the virus, many people around the world still decided to go on vacation despite the potential risks. Several studies and statistics (e.g., McKinsey & Company, 2020a, 2020b, 2020d, 2020e, 2020f, 2020g, 2020i, 2020j, 2020k, 2020m, 2020n, 2020o, 2020p; UNWTO, 2020a) showed that compared with the global level there was a surge on national travel, of people who traveled to smaller and uncrowded destinations, especially not very far from travelers' hometowns.

The same fate befell China in early 2021. In fact, on the basis of data collected from the administrations of culture and tourism, wireless telecom operators and online travel agencies, the Ministry of Culture and Tourism publicized the key data regarding domestic tourism during the Chinese New Year of 2021: this national holiday witnessed a total of 256 million domestic trips made by Chinese travelers, a year-on-year growth of 15.7 percent, which has recovered to 75.3 percent of pre-COVID-19 level during the same period of 2019. The Chinese New Year domestic trips resulted in a total of tourism-related revenue reaching up to 301.1 billion RMB yuan, registering a growth of 8.2 percent from the year before and equivalent to 58.6 percent of what had been achieved in 2019 (China Tourism Academy, 2021). It is safe to argue that in China, as elsewhere, the short trip has become the norm. Hygiene and safety were the major concern for Chinese travelers during the 2021 Chinese New Year. Various forms such as city tours, suburban excursions, road trips, family outings and Chinese New Year theme activities have been increasingly gaining popularity. As 2021 marks the 100th anniversary of the Communist Party of China, in particular, the elderly were inclined to revisit the tourist attractions bearing special historical significance. Local governments offered various vouchers and discounts on tickets to scenic spots as incentives to boost the potential consumption.

To mark the end of Chinese New Year celebrations, village and cities all over China were transformed into a sea of bright lights. The Lantern Festival came 2 weeks after Chinese New Year's Day, February 12. This was a time for family reunions and visits to crowded lantern-lighting shows and some riddle-solving games. The scenes were a stark contrast to the previous year's festivities, which were muted because of the start of the pandemic.

An international pull on Vrbo (2020) that involved over 8,000 family travelers across eight countries in the world revealed that 65 percent of respondents have a preference for domestic travel and to visit "outdoorsy" destinations rather than urban contexts. About 59 percent of families said they were more likely to drive than fly.

We can name this resistance to traveling abroad as "*abroadphobia*." Today we are certainly not facing a phobia of a pathological nature, but the anguished fear of leaving the home country could linger over time, even after the health crisis completely disappears, as a consequence of the social and cultural shock and the conditioning that the pandemic has caused (Global Data, 2020).

Conceptually, the phenomenon that I am describing under the name of "*abroadphobia*" is different from what have been defined by scholars as "*hodophobia*" and "*tourismphobia*."

The first one refers to an irrational fear of all travel (Sutherland, 1991), which has been classified as a specific anxiety disorder in the *Diagnostic and Statistical Manual of Mental Disorders* (DSM-5) and in the international classification of diseases (American Psychiatric Association, 2013). Hodophobic people experience intense and persistent fear or anxiety when they travel or they think they have to do it (Singh, Awayz and Murali, 2017).

They avoid travel if they can, or are forced to treat the acute anxiety travel causes them with antianxiety medication (Comer, 2010).

The second, "tourismphobia," which appeared 10 years ago, refers to the state of anxiety and discomfort spread among the locals when receiving a huge number of tourists which outpaces the accommodation capacity of the territory. The term was first used by the anthropologist Delgado (2008) in an article entitled "Turistofobia" in the Spanish newspaper "El País."

Since then, the term "tourismphobia," together with the concept of "overtourism," has been widely used in the literature on the subject (e.g., Postma, 2013; Bellini, Go and Pasquinelli, 2016; Milano, 2017, 2018; Koens, Postma and Papp, 2018; Milano, Novelli and Cheer, 2019) to investigate the unsustainable practices of mass tourism within urban, rural and coastal spaces and the responses that these have generated among academics, professionals and social movements.

The current circumstances, however, are still different: people want to and also feel the need to travel, so they seek a compromise between the desire for mobility and the fear of infection.

Tourism represents a way to escape from the monotony of daily routine and the restrictions imposed by the pandemic, but at the same time it is seen as the enemy, a possible way to get infected. As a result, the term "abroadphobia" fits this context. Apart from government restrictions and the closing of some borders, statistics and studies on the subject collected around the world show that people prefer to practice tourism without leaving their own countries (e.g., Corbisiero, 2020; Corbisiero and La Rocca, 2020; Dinev, 2020; MAKRO, 2020; World Travel and Tourism Council, 2020). As recently reported by UNTWO (2020b), domestic tourism showed positive signs in many markets during 2020 since people tended to travel more closely to home. A rediscovered trend reported that travelers have started going for "staycations." The term is a portmanteau of stay (meaning stay-at-home) and vacation (Oxford English Living Dictionaries, 2019) and became widely used in the United States in May 2008, when due to rising gas prices, many Americans decided to cut their travel expenses by taking a vacation close to home. During the pandemic, this made a strong comeback.

Hotels have been the main promoters of this practice, in response to the decrease in bookings by international tourists. They have reinvented themselves and have begun to prepare a series of offers specifically dedicated to their fellow citizens, such as reserved discounts or special services. During their staycations, people had the opportunity to use all the internal services of the accommodation facilities, such as saunas, swimming pools, restaurants, Internet connection services and so on, all amenities usually experienced mostly during far-from-home travels. This type of offer has allowed many people suffering from abroadphobia to be able to experience a small romantic getaway as a couple, a starry dinner, a weekend with the family to spend in a place other than their home. To the world of hotels and restaurants, which have suffered greatly from the pandemic crisis, we must therefore acknowledge the merit of having encouraged a different way of traveling without too much effort, changing the space and time of everyday life, in complete safety and according to the widespread health provisions to prevent the virus.

At the same time, traveling around the cities or towns which are near to home helps address the concern of being unable to return home in case of emergency including the unexpected cancellation of return train journeys or flights. Furthermore, in foreign holidays, even though returning to one's home country is always permitted, people know that in the event of sudden quarantine or closure of national borders, it is necessary to contact consulates for the repatriation issue, which would result in a waste of energy and money, as well as great concern. Another factor that makes outbound trips less attractive to tourists today is the fear that a mandatory quarantine could be enforced suddenly on travelers, forcing them to spend a period of isolation in a less familiar location.

In addition, it must be mentioned that travelers would be afraid if tested positive for COVID-19 before the return trip, which would result in the complete ruin of the journey and the forced reliance on a health system different from that of the home country. The health system, which should be of our concern, differs from country to country: In some it is fully liberalized or fully privatized, in others, it is accessible to the general public, in still others, diverse forms coexist. The diversification of health systems, combined with possible language barriers, poses an obstacle to the traveler when communicating with the health professionals of the destination country.

Moreover, an analysis of the contemporary social scenario also leads to the witness of another emerging phenomenon different but related to the staycation one. Indeed, nowadays among travelers from all over the world the emerging desire to "make tourism safe" have been gaining ground. In other words, travelers are continuing to engage in tourism, albeit differently than in the past in terms of places to visit, time spent, accommodation facilities, travel companions, the use of technology, and so on. Despite the difficulties characterizing the contemporary world, the will to travel still exists and is motivated by confidence in the effectiveness of the vaccine and by the hope that the infection will gradually become less widespread (e.g., Lapointe, 2020; OCSE, 2020; Vishal and Aakriti, 2020).

In this scenario, the innovation and the promotion of digital tourism with "hybrid" modes of use, supported by courageous industrial policies, are important allies to encourage tourism. Innovation is able to give a decisive boost to the tourism sector to guarantee travelers' different ways of using services, both physical and digital. Thus, as will be detailed in this text, many digital solutions are already in place in some of the smartest and most technologically advanced cities in the world. The commitment of smart tourist cities to technology, sustainability, innovation or accessibility helps not only improve the quality of travelers' tourist experiences, but also the quality of life of their inhabitants.

As for the role of new technologies in the future of tourism, new technological applications such as the Internet, mobile-based interfaces or augmented-reality systems constitute significant pillars in tourists' choices. In the pandemic era, the link between tourism and technologies would seem stronger than ever due to the acceleration of mobility that the latest generation of technological services seems to offer (e.g., Corbisiero and Monaco, 2021).

Going into more detail, we can say that the challenges that COVID-19 poses to the tourism sector are multiple, and in some respects, less predictable than other sectors with the new sensitivities induced by the concern for climate change and loss of biodiversity, which leads to questioning the relationship between human beings and the Earth. The paradigm of "make tourism safe" is based on new priorities on the part of travelers, which can be summarized as the search for: peace of mind, open spaces, health, hygiene and physical distancing.

This assertion is manifested in the choice of trip destination: travelers divert their attention to the tourist places characterized by maximum safety, defending themselves from the risks associated with the COVID-19 pandemic (e.g., EY, 2020). In particular, elements reassuring people include low tourist density, possibility of being outdoors and in contact with nature (e.g., Corbisiero and Zaccaria, 2020; DNA, 2020; Gursoy, Chi and Chi, 2020; Interface Tourism, 2020; Sigala, 2020), far from large cities (e.g., Bachimon, Eveno and Gélvez Espinel, 2020; Ioannides and Gyimóthy, 2020; VVF, 2020), although this can mean spending fewer days traveling than usual not to spend too much money (e.g., Azurite Consulting, 2020; HES-SO, 2020; Marques Santos *et al.*, 2020; Berger, 2020), which may result in less expenditure on traveling than before.

As expected, nowadays, tourists tend to be more prudent in the choice of destination between city and countryside. Urban places, in their eyes, appear to be more plagued by the pandemic because they are where COVID-19 spread faster. Furthermore, cities are also characterized by high population density and possible pollution. Several studies have hypothesized the existence of a correlation between smog and contagion, even if this has not yet been fully demonstrated (e.g., Anjum, 2020; In 't Veen, Kappen and van Schayck, 2020; Magazzino, Mele and Schneider, 2020; Mitra *et al.*, 2020; Zhou *et al.*, 2020). In this case, the media have played a central role in constructing the image of these cities as the major hotbeds for the virus (e.g., Antonelli, 2020; Cowper, 2020; Yu *et al.*, 2020) among the public. On the contrary, inland areas and small villages, as well small mountain areas, being less polluted and more unfrequented, sometimes remain free from contagion. This phenomenon helps to understand that travelers' preferences and behaviors have shifted toward domestic and regional vacations and outdoor activity and excursion, all conditions that will reign in the short term.

The ongoing vast and violent process has already re-defined the relationship between city and inland area as well as the interaction between man and nature. The new frailties, fears and needs have promptly called the tourism sector to recognize these conditions and formulate the most appropriate strategies in recovering the industry with a paradigm shift from profit to sustainability-based innovation.

In this scenario, the European Commission has also presented a package of guidelines and recommendations for tourists who do not intend to give up traveling. In particular, tourists are advised to purchase travel tickets and check in online to avoid crowds, respect physical distance when taking public transport and visiting tourist attractions; wear masks wherever they go; avoid touching buttons or handles on means of transport as much as possible and follow hygiene practices; avoid going into crowded shops and favor shopping online.

Contactless technologies have become increasingly important following the publication of several scientific studies that have reported that viral particles can survive from 48 hours up to 9 days on different surfaces, depending on the material, viral concentration, temperature and humidity (e.g., Kampf *et al.*, 2020; Mittal, Ni and Seo, 2020; Sizun, Yu and Talbot, 2020).

In particular, Chin *et al.* (2020) showed that under laboratory conditions, the virus in an infectious form could last up to 3 hours on paper, up to 24 hours on wood and textiles and up to 4 days on steel and plastic. Similarly, according to studies that were conducted by van Doremalen *et al.* (2020), at a temperature between 21 and 23 degrees Celsius and with a relative humidity of 40 percent, the infecting virus is detectable, in laboratory conditions, up to 4 hours on copper, 24 hours on cardboard, 48 hours on steel and 72 hours on plastic.

As for the retail sector, to respond to the economic crisis, but also to minimize contact among people, some stores have been converted into e-commerce hubs. For example, in February 2020, the Swedish giant H&M reorganized its network of shops. Some of them were converted into logistics hubs for deliveries, for the collection and for return of goods purchased online. Thus, the company began the assessment of the number of warehouses needed in each country. In the following months, the Spanish group Inditex and the US company The Gap Incorporated also carried out a similar initiative. At the beginning of 2021, the Spanish group El Corte Inglès decided to transform the store in the town of the autonomous community of the Basque Country Eibar into a warehouse for the sale of take-away food.

Similarly, the British brand Burberry at the end of March 2021 created an interactive virtual replica of its flagship store in Ginza, the Tokyo's tourism district, in which customers had the opportunity to buy the new collection. The virtual store was created in collaboration with Elle Digital Japan. The main feature of the digital store is that it faithfully reproduces the physical store, with three floors of commercial spaces, so as to offer customers the sensation of moving around in its different areas to see and buy products online. Until April 18, 2021, the site also hosted five short films starring actress Elaiza Ikeda. In the short films, she gave free styling tips to clients through touchpoints located throughout the virtual shop.

Evidently, the pandemic has brought out even more clearly the great resilience that tourism organizations are able to deploy when they face crises and disasters (e.g., Miller, 2008; Jiang, Ritchie and Verreynne, 2019). As we will see in the course of the manuscript, a process of transformation of routines, practices and functions has affected the entire tourism chain. The resilient character of tourism started to contribute to the construction of a new social scenario and to the definition of a new way of conceiving and practicing tourism.

The major contents and the structure of this book are listed as follows: the first chapter is dedicated to the discussion of tourism crisis management, with a focus on the measures, plans and manuals to promptly respond to disaster events in an adequate way for minimum negative impact to travelers and the tourism industry. This chapter will highlight how the sector is particularly linked to periodic moments of emergency of various kinds, which over the course of history have exposed it to more or less marked crises. In fact, the tourism sector is regularly subject to downturn due to unforeseeable events of different kinds, such as natural disasters, terrorist acts, epidemics, political and social instability. Specifically, the chapter will investigate the most widespread pandemics that have broken out in the course of contemporary history, underlining how a careful retrospect of what has happened in recent eras can help to understand what the economic and social effects of future public health emergencies will exert.

In relation to the context of COVID-19, the second chapter focuses on the solutions already been tested in some cities to address the new safety demand of travelers. The chapter intends to identify the main changes that are taking place on a global level and which may represent trends that can also be replicated in different territories. For this reason, it will consider case studies from different geographical areas and across European, American, Australian and Asian countries.

The third chapter is also based on the same assumptions. More specifically, this part reviews some of the main and most innovative solutions that, thanks to the use of new technologies, started supporting tourists in the choice and analysis of destinations, such as geographical maps to geocode various data about the destination and other COVID-19 control systems.

The focus of the fourth chapter is transportation. It is, among those already cited, the most affected in the tourism sector by the pandemic crisis. Since the beginning of the pandemic, traditional travel companies have tried to recover themselves, offering new travel experiences. After describing the main crises that the transport system went through, the chapter reviews the main and new trends taking place, reporting specific initiatives implemented in some territories.

A final dimension investigated in the book is the role and the potential of virtual travel. In the pandemic era, the link between tourism and technologies could seem stronger than ever due to the acceleration of mobility and the spread of virtual tours through online photos or videos, Google Maps, Google Street View, 3D reproductions and video games. This chapter reviews the main virtual tourism initiatives available to Internet users from all over the world. More specifically, both institutional initiatives (promoted by governments, but also by UNESCO) and those conducted by private actors are also described, such as AirBnB, which during the lockdown period launched a new service of online tourist experiences to its members.

The concluding chapter not only summarizes the main findings emerging from the previous chapters but, starting from them, also offers prospects for the future of tourism and orientations for policy-making.

Bibliography

Adey, P, Hannam, K, Sheller, M and Tyfield, D (2021) 'Pandemic (im)mobilities', *Mobilities*, 16, 1: 1–19.

American Psychiatric Association (2013) *Diagnostic and Statistical Manual of Mental Disorders* (5th ed.), Arlington: American Psychiatric Publishing.

Anjum, N A (2020) 'Good in the worst: COVID-19 restrictions and ease in global air Pollution', *Environmental Sciences.*

Antipova, T (2021) 'Coronavirus pandemic as black swan event', in T Antipova (ed.) *Integrated Science in Digital Age 2020. ICIS 2020. Lecture Notes in Networks and Systems, vol. 136*, Cham: Springer.

Antonelli, F (2020) 'Emerging aspects in technocratic politics at the time of the SARS COVID19 Crisis', *Rivista Trimestrale di Scienze dell'Amministrazione. Studi di Teoria e Ricerca Sociale*, 2: 1–20.

Azurite Consulting (2020) *COVID-19 Impact on Business Leaders, Owners and Decision Makers*, San Francisco: Azurite Consulting.

Bachimon, P, Eveno, P and Gélvez Espinel, C (2020) 'Primary and secondary place of residence, the digital link and the rise of presence', *Worldwide Hospitality and Tourism Themes*, 12, 4: 369–385.

Baker, D M (2015) 'Tourism and the health effects of infectious diseases: Are there potential risks for tourists?', *International Journal of Safety and Security in Tourism/Hospitality*, 12: 1–18.

Bellini, N, Go, F M and Pasquinelli, C (2016) 'Turismo urbano e sviluppo urbano: Note per un'agenda politica integrata del turismo in città', in N Bellini and C Pasquinelli (eds.) *Tourism in the City Towards an Integrative Agenda on Urban Tourism*, New York: Springer International Publishing.

Berger, R (2020) *COVID-19: Impacts et Rebond, Transformations Sectorielles et Implications Macroéconomiques en France*, Paris: Roland Berger.

Chin, A W H, Chu, J T S, Perera, M R A, Hui, K P Y, Yen, H, Chan, M C W, Peiris, M and Poon, L M (2020) 'Stability of SARS-CoV-2 in different environmental conditions', *The Lancet Microbe*, 1, 1: 10.

China Tourism Academy (2021) *China Tourism Economy Blue Paper No. 13.*

Colleoni, M (2010) 'Mobilità quotidiana e sistemi di trasporto', in A Magnier and G Vicarelli (eds.) *Mosaico Italia. Lo stato del Paese agli inizi del XXI secolo*, Milan: Franco Angeli.

Comer, R J (2010) *Abnormal Psychology* (7th ed.), New York: Worth Publishers.

Corbisiero, F (2020) 'Sostenere il turismo: Come il COVID-19 influenzerà il viaggio del futuro', *Fuori Luogo. Rivista di Sociologia del Territorio, Turismo, Tecnologia*, 7, 1: 69–79.

Corbisiero, F and La Rocca, R A (2020) 'Tourism on demand. A new form of urban and social demand of use after the pandemic event', *Tema. Journal of Land Use, Mobility and Environment*, 1: 91–104.

Corbisiero, F and Monaco, S (2021) 'Post-pandemic tourism resilience: Changes in Italians? Travel behavior and the possible responses of tourist cities', *Worldwide Hospitality and Tourism Themes*, 13, 3: 401–417.

Corbisiero, F and Zaccaria, A (2020) 'Turismo di prossimità', in G Nuvolati and S Spanu (eds.) *Manifesto dei Sociologi e delle Sociologhe dell'Ambiente e del Territorio sulle Città e le Aree Naturali del Dopo COVID-19*, Milan: Ledizioni.

Cowper, A (2020) 'COVID-19: Are we getting the communications right?', *BMJ*, 368: e919.

Delgado, M (2008) 'Turistofobia', *El País*, 07/12/2008. Retrieved from https://elpais.com/diario/2008/07/12/catalunya/1215824840_850215.html (accessed March 2021).

Dinev, K (2020) 'The Return of Domestic Tourism', *Ibis World*, 07/10/2020. Retrieved from https://www.ibisworld.com/industry-insider/coronavirus-insights/the-return-of-domestic-tourism/ (accessed March 2021).

DNA (2020) *La Industria Turística y el COVID 19, La opinión de la demanda: Intención de compra de productos y servicios turísticos—Nuevos hábitos de consumo turístico*, Barcelona: DNA Turismo y Ocio.

EY (2020) *Global Capital Confidence Barometer: How Do You Find Clarity in the Midst of a Crisis? Addressing the 'Now' Is Critical, but Anticipating the 'Next' and 'Beyond' Is the Optimal Response to COVID-19*, London: Ernst&Young.

Farinella, D and Simula, G (2020) 'Affrontare la pandemia COVID-19: Cronache dai pascoli', *Fuori Luogo. Rivista di Sociologia del Territorio, Turismo, Tecnologia*, 7, 1: 41–51.

Franquesa, J (2011) 'We've lost our bearings: Places, tourism and the limits of the mobility turn', *Antipode*, 43, 4: 1012–1033.

Global Data (2020) *Coronavirus (COVID-19) Executive Briefing*, New York: Global Data.

Gursoy, D, Chi, C G and Chi, O H (2020) 'COVID-19 study 2 report: Restaurant and hotel industry: Restaurant and hotel customers' sentiment analysis: Would they come back? If they would, when?' (Report No. 2), Carson College of Business, Washington State University.

Han, B A, Kramer, A M and Drake, J M (2016) 'Global patterns of zoonotic disease in mammals', *Trends Parasitol*, 32: 565–577.

Haworth, J T and Veal, A V (2004) *Work and Leisure*, Howe: Routledge.

HES-SO (2020) *Impact de la crise du COVID-19sur les habitudes de voyage: Rapport de synthèse*, Sierre: Institut Tourisme & Observatoire Valaisan du Tourisme, HES-SO Valais-Wallis.

Iaquinto, B L (2020) 'Tourist as vector: Viral mobilities of COVID-19', *Dialogues in Human Geography*, 10, 2: 174–177.

In 't Veen, J C C M, Kappen, J H and van Schayck, O C P (2020) 'Luchtverontreiniging: Een determinant voor COVID-19? (Air pollution: A determinant for COVID-19?)', *Nederlands Tijdschrift Voor Geneeskunde*, 28, 164: 51–53.

Interface Tourism (2020) *Étude prévisions de voyage post-COVID 19*, Paris: Interface Tourism.
Ioannides, D and Gyimóthy, S (2020) 'The COVID-19 crisis as an opportunity for escaping the unsustainable global tourism path', *Tourism Geographies*, 22, 3: 624–632.
Jiang, Y, Ritchie, B and Verreynne, M L (2019) 'Building tourism organizational resilience to crises and disasters: A dynamic capabilities view', *International Journal of Tourism Research*, 21, 6: 882–900.
Jones, K E, Patel, N G, Levy, M A, Storeygard, A, Balk, D, Gittleman, J L and Daszak, P (2008) 'Global trends in emerging infectious diseases', *Nature*, 451: 990–993.
Kampf, G, Todt, D, Pfaender, S and Steinmann, E (2020) 'Persistence of coronaviruses on inanimate surfaces and their inactivation with biocidal agents', *Journal of Hospital Infection*, 104, 3: 246–251.
Kaplan, C (2000) *Questions of Travel: Postmodern Discourses of Displacement* (3rd ed.), Durtham: Duke University Press.
Knobler, S A M (2006) *The Impact of Globalization on Infectious Disease Emergence and Control. Exploring the Consequences and Opportunities*, Washington: National Academies Press.
Koens, K, Postma, A and Papp, B (2018) 'L'overtourism è abusato? Comprendere l'impatto del turismo in un contesto cittadino', *Sostenibilità*, 10, 12: 4383–4398.
Lapointe, D (2020) 'Reconnecting tourism after COVID-19: The paradox of alterity in tourism Areas', *Tourism Geographies*, 22, 3: 633–638.
Lüscher, T F (2020) 'MD, FESC, COVID-19: (Mis)managing an announced black swan', *European Heart Journal*, 41, 19: 1779–1782.
Magazzino, C, Mele, M and Schneider, N (2020) 'The relationship between air pollution and COVID-19-related deaths: An application to three French cities', *Applied Energy*, 279: 115835.
MAKRO (2020) *Impacto del COVID-19 en el Sector de la Hostelería*, Madrid: Makro España.
Mancinelli, F and Palou, S (2016) 'El Turismo Como Refractor', *Quaderns*, 32: 5–28.
Marques Santos, A, Madrid, C, Haegeman, K and Rainoldi, A (2020) *Behavioural Changes in Tourism in Times of COVID-19*, Luxembourg: Office of the European Union.
Mascheroni, G (2006) 'Le Mobilità Turistiche: il Turismo Come Movimento di Persone, Luoghi, Oggetti, Immagini e Comunicazione', *Annali Italiani del Turismo Internazionale*, 1: 53–64.
Mascheroni, G (2007) 'Global nomads' network and mobile sociality: Exploring new media uses on the move', *Information, Communication & Society*, 10: 527–546.
Mazzoleni, S, Turchetti, G and Ambrosino, N (2020) 'The COVID-19 outbreak: From 'black swan' to global challenges and opportunities', *Pulmonology Journal*, 26, 3: 117–118.
McKinsey & Company (2020a) *COVID-19 Belgium Consumer Pulse 5/21–5/24/2020*, Bruxelles: McKinsey & Company.
McKinsey & Company (2020b) *COVID-19 Canada Consumer Pulse Survey 5/21–5/24*, Montréal: McKinsey & Company.
McKinsey & Company (2020c) *COVID-19 China Consumer Pulse Survey 5/5–5/11/2020*, Beijing: McKinsey & Company.

McKinsey & Company (2020d) *COVID-19 Denmark Consumer Pulse Survey 5/21–5/24/2020*, København: McKinsey & Company.
McKinsey & Company (2020e) *COVID-19 France Consumer Pulse Survey 5/21–5/24/2020*, Paris: McKinsey & Company.
McKinsey & Company (2020f) *COVID-19 Germany Consumer Pulse Survey 5/21–5/24/2020*, Berlin: McKinsey & Company.
McKinsey & Company (2020g) *COVID-19 India Consumer Pulse Survey 5/22–5/25/2020*, Mumbai: McKinsey & Company.
McKinsey & Company (2020h) *COVID-19 Indonesia Consumer Pulse Survey 5/20/2020–5/22/2020*, Jacarta: McKinsey & Company.
McKinsey & Company (2020i) *COVID-19 Italy Consumer Pulse Survey 5/21–5/24/2020*, Milan: McKinsey & Company.
McKinsey & Company (2020j) *COVID-19 South Korea Consumer Pulse Survey 5/1–5/3/2020*, Seoul: McKinsey & Company.
McKinsey & Company (2020k) *COVID-19 Spain Consumer Pulse Survey 5/21–5/24/2020*, Madrid: McKinsey & Company.
McKinsey & Company (2020l) *COVID-19 Portugal Consumer Pulse Survey 5/21–5/24/2020*, Lisboa: McKinsey & Company.
McKinsey & Company (2020m) *COVID-19 Switzerland Consumer Pulse Survey 5/21–5/24/2020*, Zürich: McKinsey & Company.
McKinsey & Company (2020n) *COVID-19 the Netherlands Consumer Pulse Survey 5/21–5/24/2020*, Amsterdam: McKinsey & Company.
McKinsey & Company (2020o) *COVID-19 United Kingdom Consumer Pulse Survey 4/30–5/3/2020*, London: McKinsey & Company.
McKinsey & Company (2020p) *COVID-19 US Consumer Pulse Survey 5/18–5/24/2020*, New York: McKinsey & Company.
Milano, C (2017) *Overtourism y Turismofobia. Tendencias globales y contextos locales*, Barcelona: Ostelea School of Tourism & Hospitality.
Milano, C (2018) 'Overtourism, malestar social y turismofobia. Un debate controvertido', *PASOS Revista de Turismo y Patrimonio Cultural*, 16, 3: 551–564.
Milano, C, Novelli, M and Cheer, J M (2019) 'Overtourism and tourismphobia: A journey through four decades of tourism development, planning and local concerns', *Tourism Planning & Development*, 16, 4: 353–357.
Miller, D S (2008) 'Disaster tourism and disaster landscape attractions after hurricane Katrina: An auto-ethnographic journey', *International Journal of Culture, Tourism and Hospitality Research*, 2, 2: 115–131.
Mitra, A, Chaudhuri, T R, Pramanick, P and Zaman, S (2020) 'Impact of COVID-19 related shutdown on atmospheric carbon dioxide level in the city of Kolkata', *Science Education*, 6, 3: 84–92.
Mittal, R, Ni, R and Seo, J H (2020) 'The flow physics of COVID-19', *Journal of Fluid Mechanics*, 894: 1–14.
Morales, L and Andreosso-O'Callaghan, B (2020) 'COVID19: Global stock markets 'black swan'', *Critical Letters in Economics & Finance*, 1, 1: 1–14.
Morse, S S, Mazet, J A, Woolhouse, M, Parrish, C R, Carroll, D, Karesh, W B, Zambrana-Torrelio, C, Lipkin, W I and Daszak, P (2012) 'Prediction and prevention of the next pandemic zoonosis', *Lancet*, 380: 1956–1965.
OCSE (2020) *OCSE Tourism Trends and Policies 2020*, Paris: Organisation for Economic Co-operation and Development Publishing.

Oxford English Living Dictionaries (2019) *Staycation*, Oxford: Oxford University Press.

Postma, A (2013) *When the Tourists Flew In: Critical Encounters in the Development of Tourism*, Groningen: University of Groningen.

Ritzer, G and Stillman, G (2001) 'The postmodern ballpark as a leisure setting: Enchantment and simulated de-McDonaldization', *Leisure Sciences*, 23, 2: 99–113.

Rojek, C (1997) 'Leisure theory: Retrospect and prospect', *Loisir et Société (Society and Leisure*, 20, 2: 383–400.

Rojek, C and Urry, J (eds.) (1997) *Touring Cultures: Transformations of Travel and Theory*, London: Routledge.

Schmeller, D S, Courchamp, F and Killeen, G (2020) 'Biodiversity loss, emerging pathogens and human health risks', *Biodiversity and Conservation*, 29, 11–12: 3095–3102.

Sigala, M (2020) 'Tourism and COVID-19: Impacts and implications for advancing and resetting industry and research', *Journal of Business Research*, 117: 312–321.

Simianer, H and Reimer, C (2021) 'COVID-19: A black swan and what animal breeding can learn from it', *Animal Frontiers*, 11, 1: 57–59.

Singh, H, Awayz, H and Murali, T (2017) 'An unusual case of phobia: Hodophobia', *International Journal of Indian Psychology*, 4, 2: 79–81.

Sizun, J, Yu, M W and Talbot, P J (2000) 'Survival of human coronaviruses 229E and OC43 in suspension and after drying on surfaces: A possible source of hospital-acquired infections', *Journal of Hospital Infection*, 46, 1: 55–60.

Sutherland, S (1991) 'Hodophobia', in *Macmillan Dictionary of Psychology*, London: Palgrave Macmillan.

Taleb, N N (2007) *The Black Swan: The Impact of the Highly Improbable*, London: Penguin.

Timothy, D and Olsen, D (2006) *Tourism, Religion and Spiritual Journeys*, Oxon: Routledge.

UNWTO (2020a) *World Tourism Barometer—Special Focus on the Impact of COVID-19, May 2020*, Madrid: World Tourism Organization.

UNWTO (2020b) *2020: A Year in Review*, Madrid: World Tourism Organization.

Urry, J (1990) *The Tourist Gaze, Leisure and Travel in Contemporary Societies*, London: Sage.

Urry, J (2000) *Sociology Beyond Societies: Mobilities for the Twenty-First Century*, London: Routledge.

Urry, J (2002) 'Mobility and proximity', *Sociology*, 36, 2: 255–274.

Urry, J (2003) *Global Complexity*, Oxford: Blackwell.

Urry, J and Sheller, M (2004) *Tourism Mobilities: Places to Play, Places in Play*, London: Routledge.

Urry, J and Sheller, M (2006) 'The new mobilities paradigm', *Environment and Planning*, 38, 2: 207–226.

van Doremalen, N, Bushmaker, T, Morris, D H, Holbrook, M G, Gamble, A and Williamson, B N *et al.* (2020) 'Aerosol and surface stability of SARS-CoV-2 as compared with SARS-CoV-1', *New England Journal of Medicine*, 382, 16: 1564–1567.

Vishal, V J and Aakriti, S (2020) 'Impact of COVID-19 pandemic on hospitality sector and it's revival post vaccine: A review', *Journal of Business and Social Science Review*, 7: 53–64.

Vrbo (2020) *Vrbo Travel Trend Report*, Washington: Expedia Group.

VVF (2020) *Les Français et les Vacances. Les Premiers Impacts du Confinement*, Paris: VVF Ingénierie.

White, N R and White, P B (2007) 'Home and away. Tourists in a connected world', *Annals of Tourism Research*, 34, 1: 88–104.

World Travel and Tourism Council (2020) *To Recovery & Beyond: The Future of Travel & Tourism in the Wake of COVID-19*, London: World Travel and Tourism Council.

Yarovaya, L, Matkovskyy, R and Jalan, A (2020) 'The effects of a black swan event (COVID-19) on herding behavior in cryptocurrency markets: Evidence from cryptocurrency USD, EUR, JPY and KRW markets', *SSRN Electronic Journal*, 4.

Yu, D, Anser, M K, Peng, M Y, Nassani, A A, Askar, S E, Zaman, K, Abdul Aziz, A R, Qazi Abro, M M and Jabor, M K (2021) 'Nationwide lockdown, population density, and financial distress brings inadequacy to manage COVID-19: Leading the services sector into the trajectory of global depression', *Healthcare*, 17, 9: 2–20.

Yu, M, Li, Z, Yu, Z, He, J and Zhou, J (2020) 'Communication related health crisis on social media: A case of COVID-19 outbreak', *Current Issues in Tourism*.

Zheng, D, Luo, Q and Ritchie, B (2021) 'Afraid to travel after COVID-19? Self-protection, coping and resilience against pandemic travel fear', *Tourism Management*, 83, 104261.

Zhou, F, Yu, T, Du, R, Fan, G, Liu, Y, Liu, Z, Xiang, J, Wang, Y, Song, B and Gu, X (2020) 'Clinical course and risk factors for mortality of adult inpatients with COVID-19 in Wuhan, China: A retrospective cohort study', *Lancet*, 395: 1054–1062.

1 Tourism, security and crisis management

Introduction

The tourism sector is regularly subject to economic crisis or downturn as a result of major forces such as natural disasters, terrorist acts, political or social turmoil and epidemics.

Therefore, when we talk about tourism, we refer to a very vulnerable phenomenon that is forced to readjust whenever travelers' safety can be threatened (Ioannides and Gyimóthy, 2020).

There are many vulnerability factors in the territories that unfortunately cause a looming "tourism crisis" (Faulkner, 2001; EHA, 2002; Handmer, 2003; Miller and Rivera, 2011; Student, Lamers and Amelung, 2020). Sonmez, Backman and Allen (1994) use the expression "tourism crisis" to describe a situation in which a negative event manages to affect tourist flows and, consequently, the progress of all tourism-related activities (Reisinger and Mavondo, 2005). In other words, destinations perceived as dangerous are deeply exposed to the possibility of suffering great economic, social and cultural damages which, starting from the lack of tourist demand, can affect the many actors present in the supply chain. It should not be overlooked that the tourism supply chain involves various components, not only accommodation, transport and excursions, but also catering, crafts, production and sale of local wares and infrastructures that support the development of tourism in a destination (Yavas, 1987; Roehl and Fesenmaier, 1992; Tsaur, Tzeng and Wang, 1997; Smith *et al.*, 2019). In this sense, the tourist offer can be described as a network of organizations involved in a series of different activities that all contribute to the construction of a destination as a "tourist proposal" (Szpilko, 2017).

The decline in tourist demand constitutes a decrease in earnings with consequent loss of job opportunities, slowdown in local development and in the income and profits for the entire community.

Research conducted over the years by numerous academics and professionals in the sector (e.g., Fischhoff, Nightingdale and Iannotta, 2001; Cavlek, 2002; Lepp and Gibson, 2003; Bongkosh and Goutam, 2012; Kwaku, 2012; Uğur and Akbıyık, 2020) has clearly emphasized that when

DOI: 10.4324/9781003195177-2

the damage suffered by a destination conditions its image, tourists tend to change their plans. The more a destination is associated with threat or danger, whether real or only perceived, the more likely travelers will consider alternative travel plans. Not surprisingly, Pizam and Mansfeld (1996: 1) argue that "security, tranquility and peace are a necessary prerequisite for prosperous tourism."

Evidently, the image of a destination and its level of safety play a fundamental role for potential tourists in their choice of holiday destination (e.g., Baloglu and McCleary, 1999; Huang and Min, 2002; Huan, Beaman and Shelby, 2004; Ichinosawa, 2006). The perception of the risk associated with a given destination perhaps exerts a far-reaching impact thanks to the contribution of the media, thereby creating a significant level of unjustified anxiety among potential travelers (e.g., Durocher, 1994; Burnett, 1998; Sonmez and Graefe, 1998; Mansfeld, 1999; Min, 2003; Bhati *et al.*, 2020; Moreno Barreneche, 2020; Russo, 2020).

The objective of this chapter is twofold: on one hand, it aims to show the resilient nature of tourism, that is, its capability to resist crisis. The historical retrospect of tourism reveals the fact that human beings never quit traveling, but readjust their mobility choices after evaluating the potential safety risks of the trip destination in times of historical or social instability.

On the other hand, after a brief historical reconstruction on the main obstacles that tourism has faced starting from the new millennium, the chapter focuses on the relationship between tourism and health.

The final section of this chapter is dedicated to a discussion of crisis management in the industry of tourism as well as the countermeasures against the new threats of COVID-19 in various tourist countries.

1.1 Tourism does not go on vacation

An analysis of historical stages of international tourism on the basis of the data provided by United Nations World Tourism Organization (UNWTO, 2020a) (see Fig. 1.1) reveals that the number of international trips has always been on the rise despite the several crises which have occurred since the new century through to the end of 2019.

Since the new millennium, in particular, the year of 2000 was marked as a year of prosperity in international tourism as it registered a year-on-year growth of 7 percent. The following year, however, saw great difficulties in the sector: the Israeli–Palestinian conflict was rekindled at the beginning of 2001 and the world stock exchanges suffered from a rapid decline, which was destined to be a far-reaching one, thereby causing an economic downturn globally. The airlines and the entire tourism industry at large, during the summer of the same year, were among the first victims of this downturn, as both experienced reduced demands (UNWTO, 2002).

The 9/11 is considered to be a watershed event in the tourism history due to its unprecedented consequences: on the morning of Tuesday,

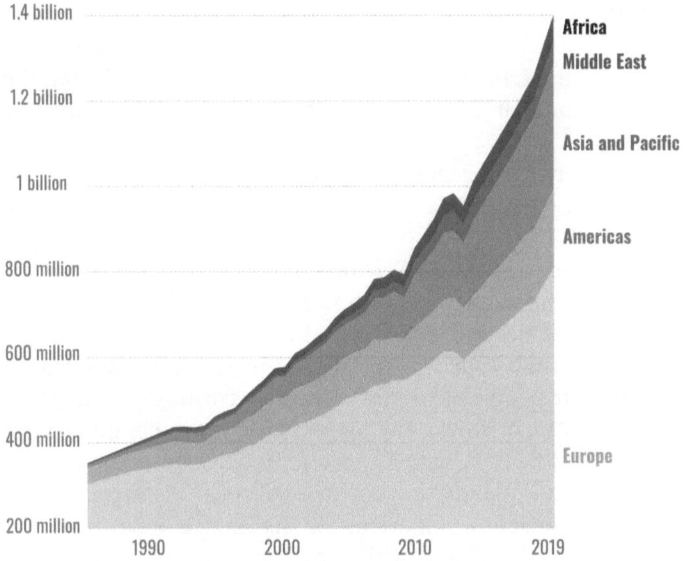

Figure 1.1 International tourism arrivals by world region.

Source: Author's elaboration based on UNWTO (2020a) data.

September 11, 2001, 19 al-Qaeda terrorists hijacked four airliners, two of which were flying between Boston and Los Angeles. These two Boeing 767s, American Airlines Flight 11 and United Airlines Flight 175, crashed into the North and South towers of World Trade Center in Manhattan. The other two airliners, American Airlines Flight 77, a Boeing 757, departed from Washington, D.C. to Los Angeles and then crashed into the Pentagon, and United Flight 73, a Boeing 757, ended itself in Pennsylvania after the passengers managed to defeat the hijackers though they intended to crash the plane in Washington, D.C. The moment when these terrorist attacks were underway, at least 2,500 aircraft were flying over the sky of the United States, all of which were later forced to land within an hour for the sake of security. Worse still, not one aircraft took off in the United States for the consecutive 4 days, which had never happened in the previous 80 years.

The 9/11 terrorist attacks deeply shook the world with a death toll of about 3,000. These victims, however, all lost their lives on a plane, one of our main means of transport. Consequently, safety and normality of travel, which lie at the root of the development of tourism, suffered a severe blow worsened by the economic crisis already underway. Under such circumstances, airlines in the world suffered most from this crisis. As a consequence, the Australian company Ansett and the American company Midway Airlines declared bankruptcy in September 2001, followed the following month by Swissair and, in November, by the Belgian company Sabena (Blake and Sinclair, 2003).

Globally, the UNWTO recorded a slack growth in tourist movements (+2.8 percent) from January to August 2001, and a significant drop (–10.9 percent) between September and December. Throughout the year, there was an overall reduction of 1.3 percent in arrivals of international travelers (UNWTO, 2001).

In particular, terrorism generated many problems in the tourism and hospitality industries on American soil, seriously impacting the economies of some states.

The year 2002 began with the war in Afghanistan, followed by the severe crisis that hit an important part of South America. Meanwhile, the stock market continued to decline, and the world economy slackened.

In this period of time, tourism experienced great uncertainty, but it continued to exist, despite people's fear of traveling. According to data provided by UNWTO (2003b), surprisingly, the biggest growth in inbound tourism was recorded in the Middle East, with Egypt and Red Sea destinations at the forefront, followed by the Pacific Coast countries of Asia. These territories gained the spotlight as principal destinations among international travelers to the detriment of the United States, which, by contrast, lost its aura as the top country to visit and became the biggest victim to the fear of terrorist attacks.

The beginning of 2003 also witnessed a war, to be specific, the conflict between the US–UK coalition and Iraq. In this same year, another incident significantly influenced the trend of tourism, namely, the epidemic called "atypical pneumonia," which led to the closure of tourist flows to East Asia. In particular, China was the most affected as it was considered the source of the contagion.

The last few years have witnessed a new shift in tourist flows: countries and territories, which were once considered as dynamic destinations, experienced a sharp decline in terms of international tourism.

However, once again global tourism data did not show any negativity (UNWTO, 2004). In other words, on a global level, people had not abandoned their desire travel; instead, they traveled with more conscious choice of destination on the basis of historical and social contingencies. The year of 2004 showed the sign of an upturn in international tourism with a growth of over 10 percent.

During that year, the demand for travel increased globally, with East Asia and China distinguishing themselves, showing the greatest prosperity in terms of inbound and outbound tourism.

In 2005, the following year, however, a series of negative events occurred, thereby significantly restraining people's desire to travel. Some of them are still alive in our memories: the Tsunami of late 2004 sweeping across South Asia, terrorist attacks striking places of great tourist importance such as London, Sharm el Sheikh and Bali, the Central American typhoons devastating New Orleans as well as a sequence of plane crashes during the summer. Despite the unfavorable global situation, international tourism showed an exponential growth in the number of global arrivals.

A social and historical analysis allows us to understand how it was possible that there was a growth in spite of the unfavorable conditions.

The reasons behind the growth were the actions implemented: drawing from past experiences, many resorts adopted various policies designed to guarantee tourists' safety. For example, Bali (e.g., Gurtner, 2016) had not experienced a dramatic rate of trip cancellation thanks to the wise publicity that had established the image of a safe island as well as promoting the determination to safeguard the locals and their guests from further danger. The Bali Hotel Association along with several other private business stakeholders obtained substantial funding to integrate security updates. In spite of all these endeavors, the local airline company Air Paradise went bankrupt.

In 2006, a new conflict emerged between Israel and Lebanon, the worsening situation in Iraq continued to worsen and increasing tension in the Middle East intensified. In face of heavy political headwinds, international tourist arrivals and profits increased by 4.2 percent, showing positivity on a global level (UNWTO, 2007).

Even in 2007, despite the tense international political landscape, tourism in various countries continued to grow at home and abroad, with a demand increasingly oriented toward the search for new, customized, nonstandardized products and tourist niches (UNWTO, 2008). In other words, on a global level, a series of differentiations was registered in motives, consumer choices and in the selection of tourist destinations (e.g., Novelli, 2005; Lew, 2008; Wilhelm Stanis and Barbieri, 2013; Pucci and Colleoni, 2016). This gradual change was made possible also thanks to the increasingly widespread diffusion of low-cost travel and online platforms that allowed the purchase of modular packages, with the possibility for customers to choose the type of accommodation, transport, transfers and other services in the destination.

From a sociological point of view, it can be argued that the synchronization of industrial society has gradually given way to de-synchronization, allowing travelers to spend their free time in an increasingly autonomous manner. Holidays and tourist resorts started to be created and constructed, broken down and recomposed (e.g., Feifer, 1985; Urry, 2003; Urry, 2007; Dujmović and Vitasović, 2015). The complexity of postmodern society has meant that tourist experiences have multiplied (e.g., Salani, 2005; Salazar, 2006; Marra and Ruspini, 2010; Marra and Ruspini, 2011; Gerosa and Magro, 2011; Dinis and Krakover, 2016; Fayos-Solà and Cooper, 2018). Companies have begun to segment the market, offering customers increasingly personalized solutions and creating real lifetime relationships, in which the buyer–seller logic has given way to that supplier–customer (e.g., Rifkin, 2000; Morgan, Pritchard and Piggott, 2002).

In 2009, however, international tourist arrivals recorded a 4 percent decline in the wake of the global economic crisis.

In the last decade, Islamic terrorism has reappeared substantially, with a series of attacks upsetting people's lives and everyday activities. These attacks were mainly carried out in Europe from Paris (2015) to London (2017), passing through Nice (2016) and Barcelona (2017) (see Fig. 1.2).

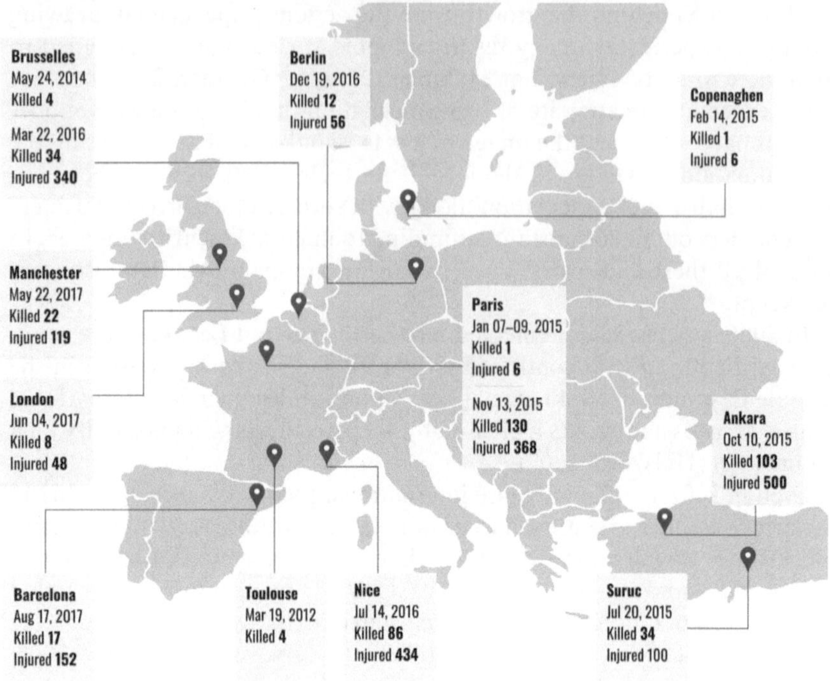

Figure 1.2 Terrorist incidents in Europe linked to ISIL (2010–2020).

Source: Author's elaboration.

The resorts affected by terrorist attacks are those that have suffered the most in terms of tourist flow and consequently, they needed a targeted and urgent intervention to prevent their tourist percentages from falling even further. However, the negative effects mainly appeared in the short-to-medium term. In the case of urban destinations, the decline was usually limited to within the days immediately following an attack (IPK, 2017). One reason for this is that the attacked cities have learned to react in a prompt and immediate manner with strict safety measures in an attempt to resume normal life and travel experiences for locals and visitors.

Recovery, if well managed, can sometimes also be seen as an opportunity to activate and improve flows for a given destination. Developing a resilience strategy for the sector is therefore essential to ensure the minimization of impacts and is particularly important in supporting companies in the sector and preparing them for a quick restart (e.g., Miller and Rivera, 2010; Cheer and Lew, 2017).

Paradoxically, in recent years, terrorist attacks have spawned the type of niche tourism in which visitors pay homage to the dead at the former attack location. The practice of visiting spots associated with death, violence and pain is known as dark tourism, such as in the case of visiting Hiroshima,

Ground Zero or Chernobyl (e.g., Seaton, 1996; Dann and Seaton, 2001; Laws and Prideaux, 2006; Singh, 2008; Sharpley and Stone, 2009; Light, 2017; Miller, Gonzalez and Hutter, 2017; Monaco and Calicchia, 2019). Thus, local policymakers, tourist industry leaders, researchers and community developers have often deployed different strategies to rebrand dark tourism sites, such as areas of rebirth to enhance destination recovery image and to promote a more disaster- and risk-resilient tourism industry.

In 2019, international tourism arrivals increased by 4 percent, with France being the most visited location in the world, followed by Spain and the United States.

1.2 Tourism and health

Evidently, the most influential factor in the tourism and travel sector is the exposure to health hazards. In particular, since the new century, the fear of travel conditioned by epidemics arose from as early as 2002, in which panic spread due to SARS, to 2019, the year of the COVID-19 breakout.

SARS (severe acute respiratory syndrome) is a viral respiratory disease caused by the SARS-CoV virus which belongs to the coronavirus family. The main symptoms include fever, cough, difficulty in breathing and, in the most severe cases, respiratory failure (Sah *et al.*, 2020). Its transmission among people occurred mainly through direct contact. This disease was first detected in November 2002, in Guangzhou, China, but was only reported by the Chinese government to the World Health Organization (WHO) in February 2003. For this reason, it was not until mid-March of the same year that the WHO recognized the disease as a real global health threat, after, in fact, the virus had already spread along the main routes of international air travel (e.g., Collins, 2005; Al-Tawfiq, Zumla and Memish, 2014).

Consequently, the Chinese government was severely criticized for its failure to inform the WHO promptly. Between 2002 and 2003, the virus infected 8,437 people worldwide and caused 813 deaths, before its disappearance in 2004 (WHO, 2007). In view of the terrible situation, for the first time in history, the WHO issued a travel advisory worldwide, that is, a series of travel-related recommendations for specific geographic areas due to the onset of an infectious disease. In particular, travelers and health workers were warned about the dangers of contagion. They were strongly recommended against nonessential travel to the destinations most affected by SARS (e.g., Clegg, 2010; Mundaca Shah, 2016).

Hong Kong was one of the cities most affected by the SARS epidemic, both in terms of the number of victims and the negative economic impact (Lo, 2006). There was a dramatic drop in the arrivals from abroad, by more than 65 percent compared to the same period in the previous year, and the occupancy rate of most hotels dropped from 79 percent to 18 percent. The pandemic of 2003 worsened the already-difficult economic situation for the mobility sector, which was just recovering from the travel blockage caused by the terrorist attacks of 9/11 (e.g., Pine and McKercher, 2004). In Singapore, tourist arrivals also dropped by more than 70 percent, forcing Singapore Airlines to send its 6,600 flight attendants and pilots on unpaid leave (e.g., Wilder-Smith, 2006).

On the contrary, the impacts of the epidemic on the Western economy were limited. The disastrous effects on tourism in the short term were only found in the Far East (Keogh-Brown, 2008).

However, most Asian economies quickly recovered thanks to the prevention strategies developed by local governments (e.g., Lee and Warner, 2007). During the epidemic, they activated forceful measures in an attempt to control the source of infection by forced isolation of the infected in government-regulated premises.

In compliance with the quarantine policies and regulations, a control system was also established at the facilities through phone calls and the installation of surveillance cameras. In addition, police controls were further tightened through the use of electronic bracelets. SARS was the first pandemic that has given rise to a specific legislation against perjury on health condition: people who lied about their health about SARS were legally punishable. In addition, the separation of infected zones being off-limit to the general public and isolation zone hospitalization of the infected was introduced (e.g., McTernan, 2004).

Such containment measures and means of international communication proved to be effective in both preventing the large-scale spread of the virus and establishing an image of health security in the Far East in the minds of travelers.

Similarly, crisis management support bodies were started in response to the lack of interconnected planning both on the governmental level and on the strictly managerial level underlying the tourism sector.

A prime example, in this sense, is the TERN (Tourism Emergency Response Network). This is a group of the world's leading tourism associations attached to the WHO, and was officially established in Washington, D.C. in April 2006. Initially, it worked as a purely advisory body to combat SARS. Over time, it has also acquired the function of drawing up concrete regional, national and global action plans for all the threats affecting tourism. To date, TERN is characterized by its independence and interdependence in which each partner shares the common goal of making travel and destinations safe for tourists. The guidelines underlying its establishment can be summarized in the following points: working closely with the United Nations system; sharing information and ideas in real time among the network partners; preparing clear, concise and geographically specific public messages; creating contacts with the media and the management of communication flows to disseminate clear and updated information.

Although it has nonregulatory power, TERN is now proposed as a valid basis for communicating and representing the interests of global tourism in times of crisis. Therefore, the committee has made all its members aware of their responsibility to help improve travelers' well-being and mitigate the impact of natural disasters (including pandemics and health emergencies) on the tourism industry.

Currently, the responsibility of TERN is under the Risk and Crisis Management (RCM) department of UNWTO. Even in noncrisis periods, TERN members are regularly invited to participate in simulation activities in order to prevent and mitigate the impact of any future catastrophes.

A few years after the establishment of TERN in 2006, the world experienced a further threat to health and tourism, the so-called swine flu, which was caused by the infection of the orthomyxovirus in pigs with particular reference to the H1N1 strain. The first human case was registered in Mexico on April 12, 2009 and within 15 days, cases were confirmed in 12 other countries. In 4 months, the H1N1 virus spread to most countries of the world causing over 284,000 deaths (Dissertori, 2017).

International travel played a central role in the spread of the virus worldwide. Suffice to say that in the sole months of March and April 2009, approximately 2.35 million passengers moved from Mexico to 164 countries by plane (Khan, 2009). To contain the contagion, many governments implemented restrictive measures on travel to and from Mexico. Similarly, the Mexican government implemented containment actions, forcing the closure of bars, clubs, restaurants and museums in Mexico City and its surrounding areas.

Clearly, Mexico recorded a net reduction in arrivals, which generated an overall crisis in the tourism sector. The most affected niches were hotels, restaurants and airlines. Hotel occupancy rates dropped from 84 percent to 30 percent (Monterrubio, 2010). In the aviation sector, the low number of passengers, the high number of cancellations and the decrease in bookings led to a 70 percent reduction in profits for national airlines (UNWTO, 2010).

To reduce the economic impact of the swine flu epidemic, the Mexican government allocated over $2,700 million to support the most affected sectors. Furthermore, when new infections began to decline, various national publicity and marketing campaigns were re-activated with the aim of re-launching the image of the main destinations in Mexico at an international level (Rassy, 2013).

In 2012, a new variant of coronavirus was identified in Saudi Arabia, called MERS-CoV, a zoonotic virus that transfers from camels to humans and thus causes MERS (Middle East respiratory syndrome).

Unlike SARS, MERS has not yet been eradicated and to date there are still some cases of infection. However, the infectivity rate of MERS is significantly lower than that of SARS. Indeed, as of January 2020, the total number of cases since 2012 amounted to 2,519, with 866 deaths (WHO, 2020a). The mortality rate reaches about 34 percent.

Some studies have highlighted that the MERS epidemic has had few significant impacts on the tourism sector. As for South Korea, for example, during the peak of the infection, tourist arrivals from China fell significantly, while on all other types of tourism from abroad, its influence was insignificant (e.g., Shi and Li, 2017; Choe, Wang and Song, 2020). Similarly, MERS had a small or even a negligible impact on tourism in Bali (Atmodjo, 2014).

1.3 Tourism management plans during times of COVID-19 emergency

Until before the COVID-19 pandemic, studies on the recovery of tourism from various crises claimed that the travel and tourism sector benefitted from an excellent resilience. In this regard, in 2019 the World Travel & Tourism Council (WTTC, 2019) highlighted that the tourism sector's ability to return to precrisis levels had shortened from 26 months in 2001 to 10 months in 2018. With specific regard to epidemics, in 2018 the time estimated to return to normality was 19.4 months on average. However, the worldwide spread of COVID-19 will have a far more devastating impact on the tourism industry than the damage caused by other health crises (OECD, 2020). In fact, the pandemic has caused a global suspension of national and international travel, causing unprecedented damage to the economy and to the tourism industry (Gössling, Scott and Hall, 2021). This means that it will take a far longer period for our lives to get back on track.

The first country that experienced a tourism crisis due to COVID-19 was China, where the infection started and then from which it spread. In the first weeks of January 2020, many airlines already started to cancel inbound and outbound flights from China. Likewise, many states deployed countermeasures preventing and closing the national borders to passengers from China, either by sea or by air. As will also be explained in the following chapters, this situation caused large losses in all sectors related to tourism, as well as transport, within a few months (e.g., Ashikul *et al.*, 2020; Newsome, 2020; Dube, Nhamo and Chikodzi, 2021).

The same situation was then found in almost all the countries of the world, as they were hit by the virus.

On the macroeconomic level, a real shock to global supplies was detected due to the fact that many nations began lockdowns when they were most affected by the virus.

A related factor that has affected the world economy is the wave of unemployment due to the closure of manufacturing and commercial activities.

The entire global tourism sector saw in 2020 a $1.3 trillion loss following the travel restrictions caused by the COVID-19 pandemic (UNWTO, 2021). This volume represents more than 11 times the loss recorded during the global economic crisis of 2009 and corresponds to a 74 percent drop in world tourist arrivals compared to 2019. Asia has been the hardest-hit continent. In 2020, international arrivals in this area decreased by 84 percent compared to the previous year. The Middle East and Africa recorded a 75 percent drop. In Europe, international arrivals fell by 70 percent, while the Americas suffered a loss of 69 percent.

The tourism sector had immediately realized that if no initiatives were adopted to make safe international travel possible, the crisis would have been far from over.

This has been also accentuated by the fact that many countries were forced to reintroduce tougher travel restrictions such as quarantines, mandatory testing and full border closures due to the evolving nature of the pandemic (e.g., McKibbin and Fernando, 2020; Więckowski, 2021).

In light of the status quo, the tourism industry is predicted not to recover prepandemic level before 2023.

According to Euromonitor (2020), starting from the end of the most acute phase of the pandemic it will take 3–5 years to see a full recovery in the travel industry. Several factors weigh heavily on the rate of recovery: first, obviously, the effectiveness of the vaccine and its administration to the entire population; furthermore, according to the WTTC, in order to curb the crisis in the tourism sector it is advisable for countries to coordinate with each other, standardizing procedures to ensure safe travel (WTTC, 2020).

In this sense, this coordination has to balance the recovery needs of the tourism industry with the health security of citizens. To find a strategy of appropriateness and consensus, UNWTO has called for close collaboration with the WHO, identifying a number of health measures necessary to minimize the impact of the virus on travel and the tourism sector in general.

In line with these considerations, during 2020, the European Union prepared a series of projects aimed at a gradual recovery of the tourism sector following the pandemic, with specific attention to issues relating to safety in travel (European Commission,2020).

Even if COVID-19 (with new variants of the virus) seems to spread in a completely different and much more sustained way than previous localized epidemics, it is safe to argue that identifying historical parallels is useful for understanding what the essential strategic aspects are in order to manage the crises generated on tourism by infectious diseases.

In fact, there are clear similarities among the syndromes of the past pandemics and the globe-sweeping COVID-19. A comparative analysis among the different events allows us to identify some fundamental points. First, general contagion management and prevention precautions are necessary to effectively reduce the risk of exposure to virus carriers and impose on travelers heading to high-risk destinations. The impact of the virus can be reduced by introducing and respecting tough measures, thus avoiding tragic human and economic costs. In this regard, history teaches us how containment and prevention efforts have been essential in limiting the impact of health crises on the local and international economy. Forced lockdown has been revived in recent times, following the positive results achieved with its implementation during the SARS epidemics in 2003 and Ebola in 2014, respectively. All countries should therefore draw lessons from the past and invest in the management of pandemic events, making this a top priority.

Moreover, immediate and clear communication is of great significance in times of crisis. Attempts to deny the problem for the fear of economic and social consequences in general, and on the tourism sector in particular, are extremely negative, as they would accelerate the spread of the virus. Such is the case of China, which underestimated the COVID-19 emergency, and failed to understand the importance of promptness and the gravity of the global impacts.

On the contrary, communication can and must be used as a strategic element to promote territories only when the state of emergency is passed. Furthermore, the WHO activity and its role in issuing global alerts is fundamental, with reference to the recommendations related to travel and international mobility too.

Then, with specific reference to the tourism sector, over time we have witnessed some imbalances in international cooperation. Crises have brought out the need to strengthen partnerships and to identify shared strategies on a global level (e.g., Bishop, 2005).

In 2007, following SARS, the International Health Regulations of the World Tourism Organization were revised, underlining the need for global interventions to avoid the spread of threats to public health even beyond national borders. In other words, the identification of collective measures is a necessity to support every country in the world to prevent the spread of viruses and save lives. In this regard, the WHO has proposed itself as the international coordinator of an integrated and effective system for addressing the threats of diseases and health disasters globally.

The close collaboration among TERN and other supranational bodies and agencies attached to the UN such as the WHO and the International Civil Aviation Organization (ICAO) is therefore of fundamental importance in strengthening the work carried out by UNWTO in preparing for crises that could negatively impact the global tourism economy.

Although in 2020 UNWTO advanced a series of recommendations to support employment and economies through travel and tourism for the sake of mitigating the socioeconomic impact of COVID-19 and accelerating recovery, these were only minimally taken into account by governments, with substantial differences between one country and another (e.g., Collins-Kreiner and Ram, 2020). National policies have often been influenced by political issues. Consequently, the virus-stricken regions of East Asia and Europe have adopted different pandemic strategies which have led to different results. A comparison between the use of masks across different countries is useful: despite the proven effectiveness of reducing the chances of infection (e.g., Fisher *et al.*, 2020), some governments (such as the British, Dutch and Finnish) initially did not recommend them. In addition, several European countries (e.g., England, the Netherlands, Sweden) upheld the idea of a "herd immunity." At an early stage, these governments allowed their citizens to be deliberately infected with the virus in order to create antibodies in the general population so that they could protect the economy and freedom of movement (e.g., Calnan, 2020; Fidler, 2020; Orlowski and Goldsmith, 2020; Randolph and Barreiro, 2020). By contrast, none of the East Asian governments have implemented such a policy for their citizens (Ho, 2020).

In such a fragmented and nonunitary context, it is clear that "abroadphobia" has begun to make its way. If every country in the world keep decide to follow its own set of rules, travelers are afraid of leaving their country because they do not know what awaits them once they cross the borders. Following the same logic, it is easy to understand why foreign people can be viewed with suspicion.

Under the current circumstances, which are still heavily influenced by the pandemic, it is essential for governments to prepare crisis management plans to restore tourism after disasters (e.g., Mitroff, 1988; Mitroff and Pearson, 1993; Peters and Pikkemaat, 2005; Moe and Pathranarakul, 2006).

Starting with the crises that have affected tourism since the new millennium, management has emerged as an area of substantial interest. In this regard, a series of strategic approaches and models have appeared (e.g., Glaesser, 2006; Huang, Tseng and Petrick, 2007).

However, research on this topic (e.g., Beirman, 2003; Ritchie, 2004; Seyondcott and Laws, 2006; Smith and Elliott, 2006) has long emphasized that it is important that destinations do not only have clear plans to implement in the case of critical events to minimize the level of damage and restore normal conditions, encouraging tourists to reconsider those territories as possible travel destinations. They should also work in advance on the preparation of plans to avoid or minimize the risk of exposure. Crisis management, in its most basic form, involves being prepared before the crisis strikes, effectively executing the crisis management plan during adverse events and quickly returning to normal after the crisis (e.g., Cuhls, 2019).

In the final analysis, crisis management refers to the planning and preparation of processes aimed at forecasting, even before managing, the negative effects of crises and disasters on tourism. More specifically, using various predictive analysis methods and tools, as well as mapping and monitoring data, the preparation of these strategies generally involves a risk assessment relating to danger and the probability that these events occur in a given territory (Martens, Feldesz and Merten, 2019). This operation is possible by a careful study of previous crises in the same territory or in other contexts with similar characteristics (e.g., UNWTO, 2003a; Yu, Stafford and Armoo, 2005; Carlsen and Liburd, 2008).

The most effective recovery strategies focus on the achievement of several results. In line with the definition of disaster recovery proposed by the International Strategy for Disaster Reduction (ISDR, 2004), the recovery process implies a holistic response to crises, which also includes economic, social and structural investments aimed at strengthening the supply chain. In concrete terms, territories experiencing a period of crisis must find the most suitable measures to recover from the damage, improve their tourist offers and prepare new marketing strategies to be more competitive on the global market and thus increase their self-sufficiency (Fabry and Zeghni, 2019). Integrated strategies to revive tourism may also include the implementation of new policies and strategies that involve the public and private sectors, including the mass media (e.g., Avraham and Ketter, 2008; Blackman and Ritchie, 2008; Dimitry, Sokolov and Dolgui, 2014).

Contrary to the recommendations made in the mid-1980s (e.g., Gee and Gain, 1986; Gartner and Shen, 1992), the most popular tourist destinations still seem to lack adequate or effective crisis management planning (e.g., Dobrowolski, 2020). This problem particularly concerns developing countries, where the expertise, resources and skills necessary to efficiently predict and deal with the onset of dangers are partially or completely absent.

Conclusions

The tourism sector has been experiencing a period of severe recession since the COVID-19 pandemic started. Even though, compared to the past, many steps forward have been made in preventing the impacts that epidemic crises can have on the tourism sector, the proposed analysis allows to argue that there are still many ongoing challenges and yet a clear vision of how to face them seems to be lacking.

Therefore, now more than ever, it is essential to learn from the history of previous infectious diseases to better address the pandemic situation and limit the negative impacts of COVID-19 on the travel and tourism sector. This operation can be useful both for identifying the mistakes to be avoided and to improve the tourism sector's management of the crisis. We have realized, for example, that compared to the past, the risk of contracting and transmitting new diseases has significantly increased in the contemporary world, also due to the multiplied possibilities of international mobility.

However, in the long run, border closures are not a viable solution to truly fight against the pandemic. The resulting economic downturn could be much worse than all the other negative effects of the infection put together. The following chapters will show that globalization and the technological revolution have provided us with many tools that could now be effectively used to pave the way for postpandemic tourism. For this reason, the desire to "travel safely" should become a constant imperative even beyond the crisis. In other words, the recommendations imposed on passengers in this specific historical period should become made into a *vademecum* to be respected in the future too, even when traveling in leisurely and unthreatening circumstances.

Likewise, people must recognize the importance of engaging in constructive dialogue with experts who can identify the best strategies for action against and containment of the infection. Similarly, as anticipated, it is essential to adopt shared strategies at a global level, since all governments must work for the pursuit of a common goal. More specifically, as regards the tourism sector there are at least three lines to follow: first, greater cooperation among health authorities, tourism institutions and operators is desirable. This must take place above all in terms of information exchange and in the definition of guidelines and crisis management plans specific to the industries belonging to the sector. Secondly, the importance of the work done by supranational bodies such as the WHO and UNTWO should be emphasized. Their recommendations are not yet fully taken into consideration by each government, and to change this, a fragmentation of policies and a territorial differentiation of their implementation should be involved. Finally, history teaches us that better management and final resolution of pandemic crisis needs transparent communication among both governments and information spreaders; that is, they should update and exchange information, thus protecting travelers in case of danger, but also reopen borders when the crisis is over. Especially in the recovery phase, the role of the media is essential to avoid uncontrolled psychosis and to rebuild trust among the public in tourism.

Bibliography

Al-Tawfiq, J A, Zumla, A and Memish, Z A (2014) 'Travel implications of emerging coronaviruses: SARS and MERS-CoV', *Travel Medicine and Infectious Disease*, 12, 5: 422–428.

Ashikul, H, Farzana, A, Mohammad, W and Ishtiaque, A (2020) 'The effect of coronavirus (COVID-19) in the tourism industry in China', *Asia Journal of Multidisciplinary Studies*, 3: 1.

Atmodjo, W (2014) 'MERS has Little Impact on Tourism', *The Jakarta Post*, 05/10.

Avraham, E and Ketter, E (2008) *Media Strategies for Marketing Places in Crises: Improving the Image of Cities, Countries, and Tourist Destinations*, Oxford: Butterworth Heinemann.

Baloglu, S and McCleary, K W (1999) 'A model of destination image formation', *Annals of Tourism Research*, 26, 4: 868–897.

Bhati, A S, Mohammadi, Z, Agarwal, M, Kamble, Z and Donough-Tan, G (2020) 'Motivating or manipulating: The influence of health-protective behaviour and media engagement on post-COVID-19 travel', *Current Issues in Tourism*, 1: 1–5.

Beirman, D (2003) 'Restoring tourism destinations in crisis: A strategic marketing approach', in R L Braithwaite and R W Braithwaite (eds.) *CAUTHE 2003: Riding the Wave of Tourism and Hospitality Research*, Lismore: Southern Cross University.

Bishop, D (2005) 'Lessons from SARS: Why the WHO must provide greater economic incentives for countries to comply with international health regulations', *Georgetown Journal of International Law*, 36: 1173–1226.

Blackman, D and Ritchie, B (2008) 'Tourism crisis management and organizational learning', *Journal of Travel & Tourism Marketing*, 23, 2–4: 45–57.

Blake, A and Sinclair, M T (2003) 'Tourism crisis management: US response to September 11', *Annals of Tourism Research*, 30, 4: 813–832.

Bongkosh, N R and Goutam, C (2012) 'Perceptions of importance and what safety is enough', *Journal of Business Research*, 65, 1: 42–50.

Burnett, J J (1998) 'A strategic approach to managing crisis', *Public Relations Review*, 24, 4: 457–488.

Calnan, M (2020) 'Health policy and controlling Covid-19 in England: Sociological insights', *Emerald Open Research*, 2: 40.

Carlsen, J and Liburd, J (2008) 'Developing a research agenda for tourism crisis management. Market recovery and communications', *Journal of Travel & Tourism Marketing*, 23, 2–4: 265–276.

Cavlek, N (2002) 'Tour operators and destination safety', *Annals of Tourism Research*, 29, 2: 478–496.

Cheer, J M and Lew, A A (2017) *Tourism, Resilience and Sustainability. Adapting to Social, Political and Economic Change*, London: Routledge.

Choe, Y, Wang, J and Song, H (2020) 'The impact of the Middle East respiratory syndrome coronavirus on inbound tourism in South Korea toward sustainable tourism', *Journal of Sustainable Tourism*, 18, 4: 1117–1133.

Clegg, C (2010) 'The aviation industry and the transmission of communicable disease: The case of H1N1 Swine Influenza', *Journal of Air and Commerce*, 75, 2: 437–467.

Collins, J (2005) 'Severe acute respiratory syndrome (SARS) and international air travel: A survey of the economic impact and international regulatory changes', Asper Review of International Business and Trade Law, 5: 43–74.

Collins-Kreiner, N C and Ram, Y (2020) 'National tourism strategies during the Covid-19 pandemic', *Annals of Tourism Research*, 10: 1–6.

Cuhls, K E (2019) 'Horizon scanning in foresight—Why horizon scanning is only a part of the game', *Futures & Foresight Science*, 2, 1: 1–21.

Dann, G M S and Seaton, A V (2001) *Slavery, Contested Heritage, and Thanatourism*, New York: Psychology Press.

Dimitry, I, Sokolov, B and Dolgui, A (2014) 'The ripple effect in supply chains: Trade-off efficiency-flexibility-resilience in disruption management', *International Journal of production Research*, 1: 2154–2172.

Dinis, A and Krakover, S (2016) 'Niche tourism in small peripheral towns: The case of Jewish heritage in Belmonte, Portugal', *Tourism Planning & Development*, 13, 3: 310–332.

Dissertori, K (2017) *Tourism Providers' Reactions to Decreased Demand Following a Crisis. The Impact of the Swine Flu on the Tourism Market: A Panel Data Approach*, Vienna: Modul Vienna University.

Dobrowolski, Z (2020) 'After COVID-19. Reorientation of crisis management in crisis', *Entrepreneurship and Sustainability Issues*, 8, 2: 799–810.

Dube, K, Nhamo, G and Chikodzi, D (2021) 'COVID-19 pandemic and prospects for recovery of the global aviation industry', *Journal of Air Transport Management*, 92: 102–122.

Dujmović, M and Vitasović, A (2015) 'Postmodern society and tourism', *Journal of Tourism and Hospitality Management*, 3, 9–10: 192–203.

Durocher, J (1994) 'Recovery marketing: What to do after a natural disaster', *Cornell Hotel and Restaurant Administration Quarterly*, 35, 2: 66–71.

EHA (2002) *Disaster and Emergencies. Definitions*, Geneva: WHO—Department of Emergency and Humanitarian Action.

Euromonitor (2020) *Accelerating Travel Innovation after Coronavirus*, London: Euromonitor International.

European Commission (2020) *Introductory Speech of the Video Conference With the Ministers of Tourism*, Brussels: European Commission.

Fabry, N and Zeghni, S (2019) *Resilience, Tourist Destinations and Governance: An Analytical Framework*, Paris: Université Paris-Est.

Faulkner, B (2001) 'Towards a framework for tourism disaster management', *Tourism Management*, 22: 135–147.

Fayos-Solà, E and Cooper, E C (eds.) (2018) *The Future of Tourism. Innovation and Sustainability*, London: Springer.

Feifer, M (1985) *Going Places*, London: Palgrave Macmillan.

Fidler, D P (2020) 'To fight a new coronavirus: The COVID-19 pandemic, political herd immunity, and global health jurisprudence', *Chinese Journal of International Law*, 19: 207–213.

Fischhoff, B, Nightingdale, E and Iannotta, J (2001) *Adolescent Risk and Vulnerability: Concepts and Measurements*, Washington: Institute of Medicines, National Research Council.

Fisher, K A, Barile, J P, Guerin, R J, Vanden Esschert, K L, Jeffers, A and Tian, L H *et al.* (2020) 'Factors associated with cloth face covering use among adults during the COVID-19 pandemic—United States, April and May 2020', *Morbidity and Mortality Weekly Report*, 69: 933–937.

Gartner, W C and Shen, J (1992) 'The impact of Tiananmen square on China's tourism image', *Journal of Travel Research*, 30, 4: 47–52.

Gee, C and Gain, C (1986) 'Coping with crises', *Travel & Tourism Analyst*, 6: 3–12.

Gerosa, M and Magro, S (2011) *Nuovi turismi. 100 alternative al classico viaggio*, Milan: Morellini.

Glaesser, D (2006) *Crisis Management in the Tourism Industry* (2nd ed.), Oxford: Butterworth, Heinemann.

Gössling, S, Scott, D and Hall, M (2021) 'Pandemics, tourism and global change: A rapid assessment of COVID-19', *Journal of Sustainable Tourism*, 29, 1: 1–20.

Gurtner, Y (2016) 'Returning to paradise: Investigating issues of tourism crisis and disaster recovery on the island of Bali', *Journal of Hospitality and Tourism Management*, 28: 11–19.

Handmer, J (2003) 'We are all vulnerable', *The Australian Journal of Emergency Management*, 18, 3: 55–60.

Ho, H K (2020) 'COVID-19 pandemic management strategies and outcomes in East Asia and the Western world: The scientific state, democratic ideology, and social behavior', *Frontiers in Sociology*, 5: 575–588.

Huan, T C, Beaman, J and Shelby, L (2004) 'No-escape natural disaster: Mitigating impacts on tourism', *Annals of Tourism Research*, 31, 2: 255–273.

Huang, J H and Min, J C H (2002) 'Earthquake devastation and recovery in tourism: The Taiwan case', *Tourism Management*, 23: 145–154.

Huang, Y C, Tseng, Y P and Petrick, J F (2007) 'Crisis management planning to restore tourism after disasters: A case study from Taiwan', *Journal of Travel & Tourism Marketing*, 23, 2–4: 203–221.

Ichinosawa, J (2006) 'Reputational disaster in Phuket: The secondary impact of the tsunami on inbound tourism', *Disaster Prevention and Management*, 15, 1: 111–123.

Ioannides, D and Gyimóthy, S (2020) 'The COVID-19 crisis as an opportunity for escaping the unsustainable global tourism path', *Tourism Geographies*, 22: 1–9.

IPK (2017) *Report on Global Travel Trends 2018*, New York: IPK International's World Travel Monitor.

ISDR (2004) *Terminology: Basic Terms of Disaster Risk Reduction. International Strategy for Disaster Reduction*, San Francisco: UN.

Keogh-Brown, M R (2008) 'The economic impact of SARS: How does the reality match the predictions?', *Health Policy*, 20: 110–120.

Khan, K (2009) 'Spread of a novel influenza a (H1N1) virus via global airline transportation', *New England Journal of Medicine*, 361: 2.

Kwaku, A B (2012) 'Tourists' views on safety and vulnerability. A study of some selected towns in Ghana', *Tourism Management*, 33, 2: 327–333.

Laws, E and Prideaux, B (2006) *Tourism Crises, Management Responses and Theoretical Insight*, Binghamton: The Haworth Hospitality Press.

Lee, G and Warner, M (2007) *The Political Economy of the SARS Epidemic: The Impact on Human Resources in East Asia*, New York: Routledge.

Lepp, A and Gibson, H (2003) 'Tourist roles, perceived risk and international tourism', *Annals of Tourism Research*, 30, 3: 606–624.

Lew, A A (2008) 'Long tail tourism: New geographies for marketing niche tourism products', *Journal of Travel & Tourism Marketing*, 25, 3–4: 409–419.

Light, D (2017) 'Progress in dark tourism and thanatourism research: An uneasy relationship with heritage tourism', *Tourism Management*, 61: 275–301.

Lo, A C (2006) 'The survival of hotels during disaster: A case study of Hong Kong in 2003', *Asia Pacific Journal of Tourism Research*, 1: 65–80.

Mansfeld, Y (1999) 'Cycles of war, terror, and peace: Determinants and management of crisis and recovery of the Israeli tourism industry', *Journal of Travel Research*, 38, 1: 30–36.

Marra, E and Ruspini, E (eds.) (2010) *Altri turismi. Viaggi, esperienze, emozioni*, Milan: Franco Angeli.

Marra, E and Ruspini, E (eds.) (2011) *Altri turismi crescono. Turismi outdoor e turismi urbani*, Milan: Franco Angeli.

Martens, H M, Feldesz, K and Merten, P (2019) 'Crisis management in tourism—A literature based approach on the proactive prediction of a crisis and the implementation of prevention measures', *Athens Journal of Tourism*, 3: 89–101.

McKibbin, W and Fernando, R (2020) *The Global Macroeconomic Impact of COVID-19: Seven Scenarios*, Canberra: Australian National University.

McTernan, B (2004) *Political Risk Yearbook: East Asia & The Pacific 2004*, Beijing: Political Risk Services.

Miller, D S and Rivera, J D (2011) *Comparative Emergency Management. Examining Global and Regional Responses to Disasters*, Boca Raton: CRC Press, Taylor & Francis Group.

Miller, D S and Rivera, J D (eds.) (2010) *Community Disaster Recovery and Resiliency. Exploring Global Opportunities and Challenges*, Boca Raton: Auerbach Publications, Taylor & Francis Group.

Miller, D S, Gonzalez, C and Hutter, M (2017) 'Phoenix tourism within dark tourism: Rebirth, rebuilding and rebranding of tourist destinations following disasters', *Worldwide Hospitality and Tourism Themes*, 9, 2: 196–215.

Min, J C H (2003) 'A study of post-disaster tourist behavior and effective marketing strategies: The case of September 21st earthquake', *Journal of Tourism Studies*, 9, 2: 141–154.

Mitroff, I I (1988) 'Crisis management: Cutting through the confusion', *Sloan Management Review*, 29, 2: 15–20.

Mitroff, I I and Pearson, C M (1993) 'From crisis prone to crisis prepared: A framework for crisis management', *Academy of Management Executive*, 7, 1: 48–59.

Moe, T L and Pathranarakul, P (2006) 'An integrated approach to natural disaster management: Public project management and its critical success factors', *Disaster Prevention and Management*, 15, 3: 396–413.

Monaco, S and Calicchia, F (2019) 'Thanatourism: la frontiera oscura del viaggiare. Il caso del Cimitero delle fontanelle', *Fuori Luogo. Rivista Di Sociologia Del Territorio, Turismo, Tecnologia*, 6, 2: 10–18.

Monterrubio, C (2010) 'Short-term economic impacts of influenza A (H1N1) and government reaction on the Mexican tourism industry: An analysis of the media', *International Journal of Tourism Policy*, 3, 1: 1–15.

Moreno Barreneche, S (2020) 'Somebody to blame. On the construction of the other in the context of the COVID-19', *Society Register*, 4, 2: 19–32.

Morgan, N, Pritchard, A and Piggott, R (2002) 'New Zealand, 100% pure. The creation of a powerful niche destination brand', *Journal of Brand Management*, 9, 4: 335–354.

Mundaca Shah, C (2016) *The Neglected Dimension of Global Security: A Framework to Counter Infectious Disease Crises*, London: Commission on a Global Health Risk Framework for the Future.

Newsome, D. (2020) 'The collapse of tourism and its impact on wildlife tourism destinations', *Journal of Tourism Futures*, 6: 1–8.

Novelli, M (ed.) (2005) *Niche Tourism: Contemporary Issues, Trends and Cases*, London: Routledge.

OECD (2020) *Rebuilding Tourism for the Future: COVID-19 Policy Responses and Recovery*, Paris: Organisation for Economic Co-operation and Development.

Orlowski, E J W and Goldsmith, D J A (2020) 'Four months into the COVID-19 pandemic, Sweden's prized herd immunity is nowhere in sight', *Journal of the Royal Society of Medicine*, 113: 292–298.

Peters, M and Pikkemaat, B (2005) 'Crisis management in alpine winter sports resorts—The 1999 avalanche disaster in Tyrol', *Journal of Travel and Tourism Marketing*, 19, 2–3: 9–20.

Pine, R and McKercher, B (2004) 'The impact of SARS on Hong Kong's tourism industry', *International Journal of Contemporary Hospitality Management*, 16, 2: 139–143.

Pizam, A and Mansfeld, Y (1996) *Tourism, Crime and International Security Issues*, Chichester: Wiley.

Pucci, P and Colleoni, M (eds.) (2016) *Understanding Mobilities for Designing Contemporary Cities*, New York: Springer.

Randolph, H E and Barreiro, L B (2020) 'Herd immunity: Understanding COVID-19', *Immunity*, 52: 737–741.

Rassy, D S R (2013) 'The economic impact of H1N1 on Mexico's tourist and pork sectors', *Health Economics*, 22: 824–834.

Reisinger, Y and Mavondo, F (2005) 'Travel anxiety and intentions to travel internationally: Implications of travel risk perception', *Journal of Travel Research*, 43: 212–225.

Rifkin, J (2000) *The Age of Access: The New Culture of Hypercapitalism, Where All of Life Is a Paid-For Experience*, New York: Tarcher Perigee, Penguin Group.

Ritchie, B W (2004) 'Chaos, crises and disasters: A strategic approach to crisis management in the tourism industry', *Tourism Management*, 25, 6: 669–683.

Roehl, W S and Fesenmaier, D R (1992) 'Risk perceptions and pleasure travel: An exploratory analysis', *Journal of Travel Research*, 30, 4: 17–26.

Russo, A (2020) 'Fake news ai tempi del Covid-19. L'uso del fact checking per contrastare l'epidemia della disinformazione', *Fuori Luogo. Rivista Di Sociologia Del Territorio, Turismo, Tecnologia*, 7, 1: 89–95.

Sah, R, Rodriguez-Morales, A J, Jha, R, Chu, D K W, Gu, H, Peiris, M, Bastola, A, Lal, B K, Ojha, H C, Rabaan, A A, Zambrano, L I, Costello, A, Morita, K, Pandey, B D and Poon, L L M (2020) 'Complete genome sequence of a 2019 novel coronavirus (SARS-CoV-2) strain isolated in Nepal', *Microbiology Resource Announcements*, 9, 1: 1–3.

Salani, M (2005) *Viaggio nel viaggio. Appunti per una sociologia del viaggio*, Rome: Meltemi.

Salazar, N B (2006) 'The anthropology of tourism in developing countries: A critical analysis of tourism cultures, powers and odentities', *Tabula Rasa*, 5: 99–128.

Seaton, A V (1996) 'Guided by the dark: From thanatopsis to thanatourism', *International Journal of Heritage Studies*, 2, 4: 234–244.

Seyondcott, N and Laws, E (2006) 'Tourism crisis and disasters: Enhancing understanding of system effects', *Journal of Travel and Tourism Marketing*, 19, 2–3: 149–158.

Sharpley, R and Stone, P (2009) *The Darker Side of Travel*, Bristol: Channel View Publications.

Shi, W and Li, K X (2017) 'Impact of unexpected events on inbound tourism demand modeling: Evidence of Middle East respiratory syndrome outbreak in South Korea', *Asia Pacific Journal of Tourism Research*, 22, 3: 344–356.

Singh, T V (ed.) (2008) *New Horizons in Tourism. Strange Experiences and Strange Practices*, Wallingford: CABI Publishing.

Smith, D and Elliott, D (2006) *Key Readings in Crisis Management Systems and Structures for Prevention and Recovery*, New York: Routledge.

Smith, K M, Machalaba, C C, Seifman, R, Feferholtz, Y and Karesh, W B (2019) 'Infectious disease and economics: The case for considering multi-sectoral impacts', *One Health*, 7: 1–6.

Sonmez, S F and Graefe, A R (1998) 'Influence of terrorism risk on foreign tourism decisions', *Annals of Tourism Research*, 25, 1: 112–144.

Sonmez, S F, Backman, S J and Allen, L R (1994) *Managing Tourism Crises: A Guidebook*, Clemson: Clemson University Press.

Student, J, Lamers, M and Amelung, B (2020) 'A dynamic vulnerability approach for tourism destinations', *Journal of Sustainable Tourism*, 28, 3: 475–496.

Szpilko, D (2017) 'Tourism supply chain—Overview of selected literature', in K Halicka and A Wasiak (eds.) *7th International Conference on Engineering, Project, and Production Management*. Warsaw: Elsevier.

Tsaur, S H, Tzeng, G H and Wang, K C (1997) 'Evaluating tourist risks from fuzzy perspectives', *Annals of Tourism Research*, 24, 4: 796–812.

Uğur, N G and Akbıyık, A (2020) 'Impacts of COVID-19 on global tourism industry: A cross-regional comparison', *Tourism Management Perspectives*, 36: 100744.

UNWTO (2001) *Tourism after 11 September 2001: Analysis, Remedial Actions and Prospects*, Madrid: World Tourism Organization.

UNWTO (2002) *World Tourism Stalls in 2001. WTO Reports*, Madrid: World Tourism Organization.

UNWTO (2003a) *Crisis Guidelines for the Tourism Industry*, Madrid: World Tourism Organization.

UNWTO (2003b) *Tourism Highlights—2003 Edition*, Madrid: World Tourism Organization.

UNWTO (2004) *Tourism Highlights—2004 Edition*, Madrid: World Tourism Organization.

UNWTO (2007) *Tourism Highlights—2007 Edition*, Madrid: World Tourism Organization.

UNWTO (2008) *Tourism Highlights—2008 Edition*, Madrid: World Tourism Organization.

UNWTO (2010) *Tourism Highlights—2010 Edition*, Madrid: World Tourism Organization.

UNWTO (2020a) *World Tourism Barometer*, Madrid: World Tourism Organization.

UNWTO (2021) *Supporting Jobs and Economies Through Travel & Tourism: A Call for Action to Mitigate the Socio-Economic Impact of COVID-19 and Accelerate Recovery*, Madrid: World Tourism Organization.

Urry, J (2003) *Global Complexity*, Oxford: Blackwell.

Urry, J (2007) *Mobilities*, London: Wiley.

WHO (2007) *SARS: How a Global Epidemic Was Stopped*, Geneve: World Health Organization.

WHO (2020a) *MERS Situation Update, January 2020*, Geneve: World Health Organization.

WHO (2020b) *Coronavirus Disease 2019 (COVID-19): Situation Report 46*, Geneve: World Health Organization.

Więckowski, M (2021) 'Will the consequences of COVID-19 trigger a redefining of the role of transport in the development of sustainable tourism?', *Sustainability*, 13, 4: 188–197.

Wilder-Smith, A (2006) 'The severe acute respiratory syndrome: Impact on travel and tourism', *Travel Medicine and Infectious Disease*, 4, 2: 53–60.

Wilhelm Stanis, S A and Barbieri, C (2013) 'Niche tourism attributes scale: A case of storm chasing', *Current Issues in Tourism*, 16, 5: 495–500.

WTTC (2019) *Crisis Preparedness, Management & Recovery*, London: World Travel & Tourism Council.

WTTC (2020) *G20's Public & Private Sector Recovery Plan*, London: World Travel & Tourism Council.

Yavas, U (1987) 'Foreign travel behavior in a growing vacation market: Implications for tourism marketers', *European Journal of Marketing*, 21, 5: 57–69.

Yu, L, Stafford, G and Armoo, A K (2005) 'A study of crisis management strategies of hotel managers in the Washington, D.C. Metro Area', *Journal of Travel and Tourism Marketing*, 19, 2–3: 91–105.

2 Innovation in post-COVID tourism

From big data to artificial intelligence and tourism rebound

Introduction

At the end of 2020, Booking.com carried out an annual survey showing travel forecasts (Booking, 2020). According to the findings, transparency and security are the two main elements that will predominantly influence travelers' choices during the post-COVID era. The study was carried out online and collected a sample of 22,000 respondents who had taken a trip in the previous 12 months or planned to do so in the following 12 months. The subjects were 1,000 each from Australia, Germany, France, Spain, Italy, Japan, China, Brazil, India, the United States, the United Kingdom, Russia, Indonesia, Colombia and South Korea; and 500 each from New Zealand, Thailand, Argentina, Belgium, Canada, Denmark, Hong Kong, Croatia, Taiwan, Mexico, the Netherlands, Sweden, Singapore and Israel.

As mentioned in the previous pages, "make tourism safe" is establishing itself as a new social paradigm, which will last for a long time. For this reason, travelers will be attentive and demanding in their search for transparent and safe experiences when they travel just like in the prepandemic period.

As for transparency, according to the study carried out by Booking.com, 74 percent of the subjects believe that accommodation facilities should clarify booking terms such as cancellation conditions, payment and refund procedures, insurance and even more. The same situation can also be generalized to the other driving forces of the tourism supply chain, such as transport companies or large catering chains. For example, the possibility of canceling a reservation without penalties or changing departure or return dates are considered as a "must" by, respectively, 46 percent and 36 percent of travelers for their future stays.

To meet these needs, Booking.com worked to ensure that its partners have maximum success in terms of pricing strategies, availability and flexibility. In fact, the company has introduced a tool that allows its customers to implement flexible conditions with just a few clicks, in order to help accommodation facilities, receive new bookings and easily manage requests for cancellations, refunds and booking changes.

DOI: 10.4324/9781003195177-3

As regards to "make tourism safe," the major points in question are greater concern among travelers about their health and the preventive measures implemented against COVID-19 in the destinations they intend to visit. In all likelihood, tourists will take more precautions during future trips: they will sanitize their hands and the surfaces they touch more frequently; spend less time socializing, and wear masks in case of need (e.g., Ivanova, Krasimirov and Stanislav, 2021; Jaipuria, Parida and Ray, 2021; Naumov, Varadzhakova and Naydenov, 2021; Roy and Sharma, 2021).

Furthermore, when planning a trip before departure, tourists will be more thorough when searching for information regarding the infection level of the virus and the corresponding countermeasures for individuals and for the public in the trip destination (e.g., Amadeus, 2020; Peluso and Pichierri, 2020; Dušek, Sagapova and Kliestik, 2021; Kaushal and Srivastava, 2021).

Consequently, they will divert their interest to destinations and structures that will pay particular attention to health, implementing clear precautionary measures and safety systems that can make tourists feel safe, also thanks to new technologies (e.g., Anguera-Torrell, Aznar-Alarcón and Vives-Perez, 2020; Singh, 2020; Anichiti *et al.*, 2021; Chan *et al.*, 2021; Minai, Raza and Segaf, 2021; Travel Tech, 2021).

Thus, today more than ever, tourist offers cannot fail to include new forms of services, thanks to the application of innovative technologies. Even before COVID-19, digital tools were considered a fundamental drive for tourism, in all phases of the journey, from booking hotels to consulting maps, reviewing websites and online booking channels or trip apps during their holidays (e.g., Kamal, 2020; Pillmayer, Scherle and Volchek, 2021). Currently destinations and operators must exploit the possibilities offered by digital technologies to propose prime solutions in response to, in an innovative way, travelers' growing need for safety. According to a study conducted by McKinsey (2020), companies in Europe have experienced a digital transformation in customer relationships within just a few months. This transformation, however, could take 3–7 years for internal digital updates, if calculated at the innovation rate of pre-COVID-19 levels.

Travel companies are therefore ready to offer advanced services, promoting the recovery of the tourism market that will be completely different in terms of flows, distribution, behaviors, power relations and players. The speed of reaction and the ability to make the most of the possibilities given by big data and artificial intelligence (AI) will determine the new assets and the future competitiveness of the various countries (e.g., Assaf and Scuderi, 2020; Babu and Justin, 2020; Fennell, 2021; Gallego and Font, 2021; Grundner and Neuhofer, 2021; Gössling, 2021).

In this regard, we can argue that prepandemic research, data and strategies have now become obsolete, inasmuch as they have lost the statistical or predictive reference values within this scenario, which is so different from the past.

In the current circumstances, a gap between the East and the West of the world is already being registered at the international level. In fact, although the COVID-19 pandemic originated in China and spread to neighboring nations, some Far East governments have proven more successful in controlling the disease than those of the West.

In the early months of 2021, when Europe and the Americas were being plagued by a new wave of the pandemic, the local authorities in these regions maintained that China, Japan, South Korea and Thailand would be quicker in returning to a normality close to the prepandemic level. Subsequently, Australia and New Zealand moved toward a situation of normality too.

In the mentioned countries, since the outbreak of COVID-19, the number of new infections has gradually begun to be limited to a few hundred cases per day (ECDC, 2021). Looking into the reasons for this success in virus control is not easy, as it is the result of an intricate compound of organizational, social and cultural factors. In Asia, for example, preventive policies, strict quarantine rules, authoritarian governments and new technologies have ensured efficiency in bringing the pandemic under control in some countries and territories.

In this situation, the decisive factor in controlling the spread of the coronavirus was the speed at which governments introduced containment and tracking measures (e.g., Amann, Sleigh and Vajena, 2020; Calvo, Deterding and Ryan, 2020; Huang, Sun and Sui, 2020; Ienca and Vayena, 2020; OECD, 2020).

However, what must be emphasized is that the Asian countries were already well prepared for the eventuality of a pandemic. Drawing from the experience of the previous SARS and MERS epidemics, the countries of the Far East had formulated (relatively fully-fledged) crisis management strategies and operational systems in anticipation (e.g., deLisle, 2004; Knobler, Mahmoud and Lemon, 2004; Bishop, 2005; Lai *et al.*, 2020).

In Europe and North America, on the contrary, what was happening in China was perceived as a negligible problem due to the fact that, with the exception of Canada, the previous SARS epidemic had caused only a few infections among Western travelers returning from the Far East. Faced with the first alarms for COVID-19, America and Europe planned above all to strengthen controls at airports, mistakenly assuming that the contagion would remain confined to Asia also this time.

Although in recent years the World Health Organization had repeatedly urged the governments of all nations to update pandemic plans and strengthen their health systems, few western governments did so.

For this reason, in Europe, with few exceptions, the program of contact tracing was implemented with delay and difficulty, thereby proving that these preventive plans were not entirely effective. Thus, poor antiepidemic results have led many Western countries to the adoption of more restrictive measures such as total lockdown, in a vain attempt to rebuild the chain of contacts and infections.

In countries and territories with desirable preventive results, however, separating the general public from contacting a virus carrier has proved to be an effective antiepidemic measure.

The use of surveillance technologies to track the pandemic has also been the subject of harsh criticism, mainly for issues related to the protection of people's privacy (e.g., Hendl, Chung and Wild, 2020; Klenk and Duijf, 2020; Martinez-Martin *et al.*, 2020; Rowe, 2020).

The authoritarian idea, which stems from Confucianism, is deeply rooted in the Asian culture. Thus, "Confucianism has often been treated as an authoritarian philosophy that exalts the absolute authority of rulers over subjects, of fathers over sons, and of husbands over wives" (Tan, 2010: 137). In particular, to tackle the virus, they have been heavily engaged in digital surveillance and were convinced that big data could have an enormous defense potential against the pandemic. Thus, the pandemic in Asia was not only fought by virologists and epidemiologists, but also by computer scientists and big data specialists. This is a perspective that the Western world finds hard to embrace.

In fact, the European tracking apps have put more emphasis on protecting the privacy of citizens, proving to be of little use for tracking infection. Therefore, although technology can function as an effective tool, we must also realize that its assistance comes at a price at the same time, just as every coin has two sides. Technologies have proved useful as regards the monitoring of people in isolation, which in East Asian countries has been very rigorous, also thanks to the use of remote "controls." These are measures that have proved to be very effective in the fight against coronavirus, but which for cultural and ethical reasons would have been difficult to accept for Western citizens.

Finally, technology has proved to be an essential support for the practice of the rapid saliva and the antibody tests.

In general, we can argue that innovation and technologies, supported by incentive industrial policies, can be an important impetus to restarting tourism. We live in a world where the digital revolution, with its multiple applications, permeates every aspect of our society, becoming a dominant language. Data are the key elements of our times, since they represent a real resource, and create structures for collection. As a result, processing and sharing have become essential in giving immediate responses to any event of exceptional importance such as pandemics.

This chapter presents the main digital solutions that are already in place in some of the smartest and most technologically advanced countries in the world. The commitment of smart tourist cities to technology, sustainability, innovation or accessibility means not only an improvement in the quality of travelers' tourist experiences, but also in the quality of life of the local inhabitants.

2.1 It all started in China…

In December 2019, in Wuhan, China, some workers in a wet market—where even live animals are on sale—were suffering from a strange pneumonia of unknown cause. In the first weeks of January 2020, scientists identified that the subjects in question were infected with a new coronavirus, designated as SARS-CoV-2, identified as a new coronavirus mutation capable of adapting to the host organism, resulting in rapid contagion within the population. Within 2 months, the virus quickly spread to the entire city of Wuhan and its surrounding areas as well as cities.

On the evening of January 20, 2020, Zhong Nanshan, the country's most famous respiratory expert and key figure in China's anti-COVID-19 effort, confirmed the human-to-human transmission nature of coronavirus on the CCTV national television channel to the whole population. He had also fought at the forefront when China suffered from SARS in the year of 2003, so people firmly believed in him.

Meanwhile, the city of Wuhan prepared for isolation and the closure of all of activities in an effort to prevent the spread of the virus among its inhabitants. This decision was made also considering the special geographical location of Wuhan in the province of Hubei, of which Wuhan is the provincial capital. It functions as the transportation hub for central China, which adds difficulty to the prevention of disease.

Thus, on January 22, 2020, the Chinese government declared the total lockdown of the entire city would be implemented from 10 o'clock of the following day among its 11 million citizens.

This event happened in a very special period for Chinese culture and its tourism industry. In fact, it was a very short time before the Chinese New Year, which fell on January 25, 2020. Therefore, it was a very difficult decision to make at that time, not just "restrictive": many people were already planning to go back to their hometowns, and had already booked their tickets. Just like at Christmas in the West.

It was the most severe restrictive measure ever adopted in the history of China: in the city of Wuhan, airlines, trains, buses and ships were suspended, and private vehicles and going out were forbidden, except for health issues and daily purchasing (of essential goods) (e.g., Ashikul *et al.*, 2020).

Despite these measures, the virus spread in a short time in the areas adjacent to Wuhan, forcing the WHO to declare China as a "Very high risk area" on January 26, 2020, and the rest of the world as a "High risk area" (WHO, 2020a). From 20 to 30 January, the number of infected people in China was increasing every day, amounting to thousands of positive cases and nearly 5,000 deaths. After a short time, the first cases of infection were also discovered and confirmed in Japan and Mongolia and, at the end of January, the pandemic also spread to Europe, America, Africa, Oceania and the rest of Asia as well.

Consequently, on January 30, 2020, the Director General of the World Health Organization declared a state of alert for an international public health emergency (WHO, 2020b). The main measure was the recommendation to the world population to block and cancel any trips or visits to China. In other words, Chinese tourism was immediately put on hold.

Meanwhile, a WHO task force, made up of 25 scientists and researchers from China, Japan, Korea, Nigeria, Russia, Singapore and the United States immediately took action to formulate the most suitable strategies to lessen the disease's intensity. The action plan also focused on the application of big data and AI in attempts to strengthen contact tracing and offer clinical support and information as quickly as possible. In this scenario, the innovative use of technologies, medical platforms and health assistance helped to promptly address the health emergency, allowing containment and control of the infection.

The use of these new measures, still in progress, has allowed China to exit from the list of countries most affected by the pandemic.

In particular, tracking technology in China was incorporated into the popular Alipay and WeChat, two apps used on a daily basis (see Fig. 2.1).

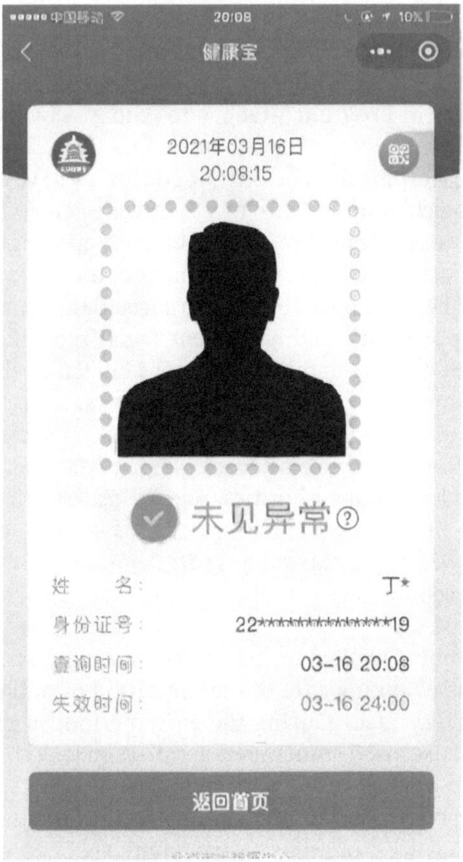

Figure 2.1 Health QR code on WeChat.

Source: WeChat app (accessed March 16, 2021).

Thus, the government's strategy was not to create a separate app with the sole function of tracking contacts, with the risk that it was neither downloaded nor used. The Alipay Health Code was created by a subsidiary of the e-commerce giant Alibaba Group. This app generates for each registered user a quick response (QR) code of a different color: green, yellow or red based on user's health data and travel history, helping to track if the person has ever travelled to areas with confirmed positive cases. The software is in use in hundreds of cities across China, thus determining whether people can go to work or use public transport. In the case of a green QR, people can move almost freely, while always scanning the code at the entrance of places such as condominiums, offices or shopping centers. If the QR code is yellow, it means there may be a danger of slight exposure to the virus (the person has travelled to risk areas). In some cases, this risk requires a week of self-quarantine. If the QR code appears red, it means that the person has come into contact with the virus. In this case, the person is obliged to remain in quarantine.

The local government of Hangzhou, a metropolis of over 10 million inhabitants in South China, with the support of the company Ant Financial, launched another mini-program attached to Alipay, which became popular in many other cities.

If people decide to apply for a health QR code, they have to do a face scan.

These technological innovations together with other digital tools had already been in place for some time for the collection of big data. As Byung-Chul Han (2017) notes, in China every moment of daily life is subject to digital observation. There are over 200 million surveillance cameras, and many of them are equipped with highly efficient facial recognition techniques. These cameras can observe and evaluate every citizen's behaviors in public places such as, shops, streets, stations and airports. Every click, purchase, contact, activity on social networks is detected. Such social surveillance is possible because there is an unlimited exchange of data across the Internet, mobile phone providers and authorities. According to the local news, people were surprised to find that certain areas of the city were off-limit to them even if they were wearing a mask thanks to the power of facial recognition technology that even manages to discover the identity of subjects through the obstacle of masks.

In some cities robots walked on the streets reminding people of the necessary precautions or taking care of the sanitation of public spaces, relieving staff from a risky task. During the early period, in some rural areas in China, people also used unmanned aerial vehicles (UAVs) as a tool of surveillance.

These robots have played a role similar to that of drones that detect people who are not wearing masks and invite them to cover their nose and mouth, or make appeals and information on good behavior using thermal cameras to locate people on the street with an abnormal body temperature.

The entrance to public places such as subways or shopping malls are mostly equipped with thermal imagers that measure the temperatures of the pedestrians in a noncontact way.

These infrastructures for digital surveillance have proven effective in containing the pandemic. For example, starting from the beginning of the state of emergency, Beijing arriving train passengers' body temperatures were automatically captured by thermal cameras. Passengers with a temperature detected above the normal range were quickly met by staff, fully equipped with protective suits and visors, to ask the person to remain in a special room that can be found in every train and subway station. A negative pressure isolation ambulance would take the person to the hospital and ask him or her to do the nucleic acid test and remain in a hospital under observation to see if he or she was confirmed as a positive case. In the event of a positive case, authorities began an epidemiological survey, using big data, to contact trace the passenger's interactions. Additionally, special staff would then contact these potentially infected people or ask community residents' committees to do so. Then the latter would send their staff to install an alarm or seal the main entrance of the house of those suspected of infection, who must undertake a home quarantine. Alternatively, suspected cases could be transported to an assembly site (usually a hotel where people with the possibility of infection were gathered) for a quarantine, where they were given daily necessities and would undergo regular nucleic acid testing by volunteers from the committee.

As for tourists, if an abnormal temperature was detected, valid nucleic acid tests and antibody tests were needed. The identified people were immediately transported by bus from the hub to a quarantine site. Tourists were forced to spend self-financed quarantine in a hotel. This means that all expenses were on tourists.

Furthermore, when the pandemic exploded in Wuhan, thousands of digital investigation teams were formed to search for potentially infected people, based solely on technical data.

Also in China, Alibaba Damo Academy, an Alibaba research spin-off, has developed a new diagnosis system based on AI that is able to detect new cases of coronavirus through computed tomographic scans with an accuracy rate of up to 96 percent, reducing the waiting times dramatically compared with that of traditional swab tests.

This new diagnostic tool was first introduced in Qiboshan hospital (specially designed to receive confirmed cases) in Zhengzhou, Henan province, as early as March 2020, and was then adopted in the following months by another 100 hospitals in Hubei, Guangdong and Anhui.

This is not Alibaba's first attempt to use AI to fight against the coronavirus. Researchers of the Damo Academy, in fact, had already developed an application for the public health service, providing information relating to SARS-CoV-2, and answering questions about the epidemic and its symptoms.

In Singapore, following the many epidemics that had hit the city, in 1913 the National Center for Infectious Diseases was established. This is an institution that has not lost strength or capacity over the years, using the most advanced technologies and remaining vigilant over control and warning systems. In particular, after the spread of the coronavirus, the city health authorities tested all cases of flu and pneumonia, quarantined the infected people and activated a remote monitoring system through an application aimed at reassuring and at the same time observing the people forced to stay at home (e.g., Liang, 2020; Sun *et al.*, 2021).

The good results obtained in the year of 2020 allowed Chinese people to celebrate Chinese New Year in February 2021. This was a great event, which involved the population (and many domestic tourists) to a high degree. In these circumstances, innovative products and services created brand new tourist experiences: 5G, augmented reality, virtual reality, AI and UAVs used for the promotion of online activities and live events. Online entertainment and trips gave tourists a new experience of immersive travel.

For example, a mini program attached to WeChat offered people a 360-degree view of the historical Dunhuang caves with the help of virtual reality technology without involving the trouble of traveling thousands of kilometers. For the first time, the National Opera Theater made its classic repertoire available online thanks to 5G + 4K technologies. The cultural atmosphere during the 2021 Chinese New Year was very strong. According to data from the Ministry of Transportation, local public libraries of all levels received 4.32 million readers and cultural centers 2.11 million visitors.

In early 2021, China Tourism Academy (2021) estimated that a total of 4.1 billion domestic trips would be completed by the end of 2021, which would have generated a total of tourism-related revenue of up to 3.3 trillion RMB yuan, registering a year-on-year growth of 42 percent and 48 percent, respectively. If the vaccination campaign is fluid and effective, the inbound and outbound market could restart at full capacity in the short-to-medium term. In the same period, some experts predicted that China would recover 30 percent of its prepandemic level over the course of the year, with inbound tourists increasing by more than 50 percent. If the situation remains stable, the same experts have speculated that revenue related to international tourism will increase by 65 percent. They also estimated that outbound trips made by Chinese travelers are expected to grow by up to 70 percent by the end of 2022.

2.2 Taiwan does not stand by

Similarly, since the start of the pandemic, the Taiwanese government used a system to send all citizens an SMS in real time to locate individuals who have had contact with infected people or to inform its population about places and buildings where people have been infected. From the earliest moments the news about the COVID-19 spread, Taiwan began using a data connection to identify potentially infected people based on travel records. At Taiwan airports, for example, arrivals from Wuhan underwent physical health checks before the possibility of human-to-human transmission of the virus was confirmed on January 20. As early as February 1, 2020, Taiwan, Hong Kong SAR and Singapore had proactively implemented travel restrictions on passengers coming from mainland China, contrary to the absence of such restrictions in other countries (e.g., Wang, Ng and Brook, 2020; Wu, Chen and Chan, 2020).

Paradoxically, Taiwan's continued friction with mainland China has favored Taiwanese success in the battle against COVID-19. In fact, Beijing had already banned individual tourism in 2019 and heavily cut group tourism on Taiwan, thus reducing the possibility of contagion. Furthermore, China has always imposed its veto to exclude Taiwan from the World Health Organization (Deng, 2020). Taiwan was thus able to act on its own, without waiting for the belated indications from Geneva on how to protect itself.

Taiwan also adopted a policy on February 6 stating that every Taiwanese could purchase a certain number of adult and child masks per week from pharmacies and clinics for $0.17 each. To allow for easier distribution and prevent long queues outside pharmacies, the Taiwanese public could order them online and collect them (later in offline physical stores.)

To ensure coordination, Taiwan has established a unified command center, led by the Ministry of Health and Welfare, which manages resources, holds daily briefings and monitors public communication (e.g., Lin *et al.*, 2020; Steinbrook, 2020). The authorities have also moved quickly to trace the infected people and map cases to identify the sources of the infection. Important campaigns have been launched to educate the population about the risks of the disease and the precautions to be taken through warnings and television commercials.

As a major technological power, Taiwan has also used all the tools at its disposal to combat the pandemic. In particular, Taiwan health insurance and immigration agencies have integrated the travel history of the previous 14 days of local and foreign residents with their health card data, allowing hospitals, clinics and pharmacies to gain immediate access to important information (e.g., Summers *et al.*, 2020).

2.3 The uniqueness of Japan

The history of how Japan has fought against epidemics informs us that incredibly far-reaching plagues have had a strong influence on the political, institutional and economic developments of the country.

In its history, Japan has been stricken by several noteworthy epidemics, with significant social, economic and religious repercussions throughout the country. Influenzas, measles, dysentery, then tuberculosis have claimed millions of victims, profoundly changing Japanese people's world views.

The worst pestilence that swept through Japan is what has been called the "minister of death" of antiquity, smallpox (e.g., Jannetta, 2014). Primarily, the outbreak of 735–737, known as the Great Smallpox Outbreak of the Tenpyō Era, left a death toll equivalent to a third of the then Japanese population. The royal family was also hit and many members of the Fujiwara dynasty, who were in power at the time, died (e.g., Farris, 1985; Hopkins, 2002).

Under that circumstances, the emperor Shomu wanted to put Japan under the protection of Buddha and had the Todaiji temple built in the capital, Nara, and, inside the Daibutsu temple grounds, the Great bronze Buddha was erected. Smallpox was demonized with the name of Hoshoshin, embodying the figure of a demon returned to earth to take revenge for wrongs suffered. To try to appease his anger, traditional rites and dances were practiced (e.g., Bamforth, 2006).

The other crushing epidemic that hit Japan was the Spanish flu of 1918. Estimated deaths amounted to over 300,000, although there are studies that speak of two million deaths (e.g., Kawana *et al.*, 2007; Chandra, 2013). The Spanish flu was an H1N1-type virus, not so dissimilar to the coronavirus. The measures taken at the time were more or less the same as those of today: stay at home as much as possible, wear masks and avoid crowded places.

Today, thanks also to the help of new technologies, Japan can boast of having fought the virus even more effectively. In fact, deaths in Japan surprisingly decreased slightly in 2020 compared with the previous year, for the first time in the previous 11 years despite the attack of coronavirus. Data provided at the beginning of 2021 by the Japanese Ministry of Health show that 1,384,544 people died in Japan in 2020, almost 9,400 fewer than the previous year (Japan Ministry of Health, 2021). This result is in contrast to what has happened in many other countries in the world, especially if we think that Japan has some characteristics that make it particularly vulnerable to COVID-19. For example, it has the oldest population in the world (ILCJ, 2017). The case of Japan shows how the prevention measures adopted there against the coronavirus have been successful not only in preventing deaths from COVID-19, but also those from seasonal flu and others respiratory diseases.

One of the reasons that could account for the decreased death toll in Japan in 2020 is the fact that wearing masks there had already been quite widespread and masks were in adequate supply even before the pandemic, as in other Asian countries. The government had also focused on a communication campaign that encouraged people to avoid enclosed spaces, crowded places and close contacts.

Furthermore, the Japanese population has been among the quickest to embrace technology in all its forms compared to many other nations. Even before the lockdown, e-commerce accounted for more than half of total retail sales (e.g., Watanabe and Omori, 2020). Technological progress also included contactless payment, a key factor in today's post-COVID world to avoid the circulation of potentially infected banknotes. The pandemic has also spurred the use of digital and robotic technology in other areas. For example, the shift to remote diagnosis and medical technology has been accelerated (e.g., Akiyama, 2020; Kono *et al.*, 2021). In addition, some Japanese electronics manufacturers have developed systems to visually suggest to people the distance to keep from others in crowded places, such as stations or shopping centers. The following are some concrete examples: Hitachi has created a device that detects the presence of subjects through sensors and draws a circle with a diameter of 2 meters around them, with fish swimming around the circumference. If two people are inside the same circle, that is, they are at a dangerous distance, the circle then changes color and the fish swim out of it. A similar device, manufactured by Mitsubishi Electric, already entered the market in April 2020. This projects a digitized animation on the floor to remind people if there is a crowd in the place they are about to enter or allows the entrance of small groups in the elevators. Moreover, Panasonic has implemented equipment that warns people of a dangerous crowd in a space by changing the color of the lights there.

Additionally, in June 2020, the Japanese Ministry of Health released a contact-tracing application called Cocoa (COVID-19 Contact Confirming Application), which is based on decentralized technology made available by the joint efforts of Apple and Google.

Japanese software uses the Bluetooth signal to detect the proximity of two smart phones. Each phone is assigned random and temporary codes for the protection of user privacy. People who test positive have to report their state of infection on the app. Consequently, after updating the state of infection, anyone who comes in contact with them is notified and has to contact the public health system.

2.4 The young South Korea

The idea that digitization and big data can be considered as the key to success in the fight against COVID-19 is also found in South Korea. At the beginning of the epidemic the situation seemed to get out of hand: the virus-infection curve increased even though the cases were concentrated mainly in large areas of the cities of Seoul and Daegu (e.g., Cooper, Mondal and Antonopoulos, 2020; Park, Choi and Ko, 2020). Subsequently, massive population tracking policies implemented through apps were adopted, including other data collected through records of credit cards or images taken from video cameras in public places. This is a policy that has taken effect since 2015. Despite some legal perplexities, it was originally formulated to deal with the MERS epidemic, which broke out in South Korea that year.

The South Korean government has also shared the data of travelers arriving from China with clinics and hospitals since the early months of 2020, verifying if any of their patients had come from that country and thus at great risk of COVID-19 infection.

To avoid overcrowding hospitals and at the same time isolating the infected, the Korean Government entered into several collaborations with alternative structures, such as hotels and conference centers, which immediately began to host the asymptomatic carriers. In the Korean model of contact tracing, intelligent technologies have played a fundamental role. The Korean Ministry of Health has hired some "trackers" whose main task has been to go through video footage against the clock, keeping a record of the target's movement and then identifying the close contacts.

The data detected by the video surveillance camera coincide with the history of the credit cards of the infected and the movements registered by their mobile phones through a retroactive global positioning system (GPS) verification, making it possible to observe and control all suspected cases. The collected data is then published on government websites and disseminated with the help of message systems and applications that reach people who have come into contact with an infected person, or who have frequented the same places. The people who receive this communication are called to receive a health test in turn. Extremely private data, such as the name or photo of the infected person, are anonymized. The only information disseminated, is related to age, origin and places visited. Once the infected person can be discharged from hospital and the places they have visited have been disinfected, the information will be removed from the shared system.

We must specify that South Korea has a very young population, with only 14 percent of the total of population aged over 65. This certainly had a weight both on the lower incidence of virus-positive cases and on the greater involvement of citizens in the monitoring of the pandemic through digital tools (e.g., Lee, Heo and Seo, 2020; Lee and Lee, 2020).

In the so-called information society (Masuda, 1981; Lyon, 1988; Castells, 1996, Castells, 2000; Lupton, 2015) young generations use the Internet and new technologies on a daily basis (Veen and Vrakking, 2006; Gasser and Palfrey, 2008; Gargi and Maitri, 2015). They are certainly more familiar than the old with embracing new forms of communication and new digital devices. They grew up with Internet and socialized under its influence. For them, the so-called networked society (Castells, 2005) is the only world they know. For this reason, from Millennials onward, people have naturally integrated technology into their lives, connecting virtual reality with real virtuality, and creating personalized use with specific needs.

Today's young generations are more informed, more mobile and more adventurous than ever (Horak and Weber, 2000). Technology plays a key role in the lives of younger generations who are now tech-savvy, hyperconnected and addicted to the Internet (Skinner, Sarpong and White, 2018; CBI, 2019).

To make up for the lack of a prepared emergency health system, South Koreans have therefore accepted technology in the control and prevention of infections, exchanging information regarding their safety and health and accepting a little less privacy in exchange for a greater control of the virus.

Moreover, COVID-19 patients in South Korea were subjected, in most cases, to home quarantine under the control of telemedicine tools maneuvered by epidemiological specialists. Also in this case, GPS tracks from cell phones, credit card data and closed-circuit cameras were used to trace the contact map of infected patients. The databases of law enforcement agencies, telephone companies, health insurance companies and financial authorities have been integrated to work together.

In addition, the "Corona 100m" app was even released, which reports both nationally and internationally the presence of a COVID virus carrier within 100 meters.

Other apps that work with Google Maps, among the most downloaded from Google Play, are "Corona Map" and "Shincheonji Location Notification," which use a collaborative system making it possible to send reports on potentially dangerous subjects or places.

A further technological contribution is represented by the automaton created by the SK Telecom company in collaboration with Omron, which is an AI robot that helps South Koreans to comply with the health regulations imposed by the government. In particular, once the robot notices a crowding ahead, it will ask people to disperse and remind them of wearing masks. It also measures the body temperature of people and triggers an alarm if this is above 37.5 degrees. To protect people's privacy, the developers have introduced a function in the robot that allows the faces of those who appear on the screen to be hidden.

The first robot was located at the SK Telecom headquarters in Seoul and was also equipped with dispenser of hand sanitizer and floor disinfection cleanser. The automaton, in fact, has ultraviolet ray lamps and disinfectant sprays that allows it to sanitize 99 percent of an area of 33 square meters in 10 minutes.

In addition, to meet the security needs of citizens and younger travelers, South Korea began opening more and more fully automated, and cashless stores during 2020. These technological stores represent a revolution that could globally characterize the world of commerce in the near future, like the self-service store Amazon Go.

This kind of shop has a concept similar to that of the Amazon Go store. With the first shop in Seattle, Amazon also chose London as the first European city to establish the new type of store. In it, the "Just Walk Out" technology automatically detects when products are picked up or put back on the shelves. Purchasing behavior is tracked through a virtual cart thanks to computerized vision, sensory fusion and deep learning. When the shopping is done, people can simply leave the shop without manual payments. In fact, shortly after leaving, they receive an electronic invoice and the corresponding amount is debited from their Amazon account automatically.

Another characteristic of contemporary contactless South Korea is the self-service stores, where customers can buy the desired products by scanning a series of QR codes with their phone. Payments are then made through the digital wallet connected to their smart phones, which has a series of sophisticated sensors to prevent theft, and immediately connect the products selected to the customer's identity.

These self-service stores are currently mainly concentrated in Seoul and are equipped with a microwave oven to heat the food purchased, a freezer, a charging station for smart phones and milk and tea dispensers.

2.5 What changes occurred from East to West?

A real-time representation of the reality implemented to intervene strategically on the evolution of an epidemic has already been tested. One example in the recent decade will suffice to illustrate this: back in 2014, big data in Africa made it possible to predict the direction of expansion of the Ebola epidemic toward Senegal, comparing real-time data on hospitalizations with those on flight routes, relevant topics on social networks and mobile phone calls (e.g., Gaudin, 2014; Tom-Aba *et al.*, 2015; West, 2015).

Therefore, the first 15 years of the new millennium witnessed a real epistemological revolution based on the possibility of detecting and analyzing a large amount of digital data and information (e.g., Crawford, Miltner and Gray, 2019; Resnyansky, 2019).

Since the start of the pandemic, dozens of countries have introduced the use of digital apps in an attempt to identify the people exposed to COVID-19, thus preventing the transmission of the infections. If in the past studies on this were accompanied by doubts, those now carried out in different countries show increasing evidence that these apps can actually help prevent infections and turn the situation back to normal, if used wisely (e.g., Kreps and Neuhauser, 2010; Golten Jutel, 2011). Contact tracing apps are a valuable tool for public health, so their integration into the local health system should be improved (e.g., Boulos *et al.*, 2014; Lee *et al.*, 2018; Walrave, Waeterloos and Ponnet, 2020).

Another prime case of this is that of Israel. In 2020, Shin Bet started monitoring the movements of COVID-19 carriers on the basis of a database originally designed in 2002 to fight against terrorism. The large amount of data in relation to travel and times of all cell phone owners were connected to Israel's telephone network under the authorized observation from the Knesset, the Israeli parliament. Within a 30-day emergency period, even asymptomatic suspected cases had to undergo quarantine.

This strategy proved to be successful: the targeted isolations, combined with the successful vaccination campaign, allowed the country to reopen theaters, gyms and tourist sites to 75 percent of their capacity as early as March 2021 (even though in the first phase these places were made accessible only to those who possess a Green Passport, a 6-month certification of vaccination).

On the strength of this milestone, in early 2021 Israel and Greece entered into an agreement that vaccinated tourists could travel freely between the two countries without imposed quarantine. This does not necessarily meant that the unvaccinated population were forbidden to travel; it only meant that those who had received two doses of the COVID-19 vaccine were exempt from self-isolation.

However, as is known to all, in the Western world, privacy is perceived differently than in Asia, or even in the most open-minded Asian countries such as South Korea and Japan, where a strong Confucian sense of community prevails.

For this reason, many countries in Europe and the United States have preferred a less technological way of flattening the curve of the pandemic, namely resorting to lockdowns.

Differently from Asian countries, the apps used in the United States and in the major European countries have been designed to ensure respect for privacy, using mainly Bluetooth or other sensors. Furthermore, their use has been on a voluntary basis and the identity of people remained anonymous for the sake of data protection. Thus, from a technological point of view, these apps in most cases stored locally, on the device, all the Bluetooth codes of other devices equipped with the same app (whether these were smartphones, smart watches or stand-alone devices such as bracelets). Encryption and anonymization systems prevented the code from being associated with the identity of the owner of that device.

The server was able to calculate for each of these codes the risk of contagion (proximity, contact time) and then to ensure that a notification could arrive to all the devices of people at potential risk through the app. The notification is a message, determined by the local health authorities, asking the general public to follow safety regulations (ranging from isolation to the obligatory antigenic swab test).

Even though in the Western world contact tracing strategies have failed to reach the desirable results compared to what have been achieved in the East, several territories have taken steps to implement technological solutions, improving the livelihood of their inhabitants and encouraging travelers from all over the world to choose them as destinations.

It must be added that in May 2020, at the end of the first wave of COVID-19 that affected many European States, step-by-step guidelines and recommendations were needed for safety conditions within the European Union to restart tourism. These were drafted by the European Commission in the Communication "Tourism and transport in 2020 and beyond" (European Commission, 2020a), with the aim of promoting domestic tourism within each European country and relaunching the tourism industry based on the principles of sustainability and, above all, of digital transition, which is considered to be one of the main tools to promote the recovery of mobility and fight the pandemic.

Examples of actions consistent with this specific objective are: initiatives of integrated information management; the promotion and marketing of the tourist offer and the interoperability among tourism portals of the different destination levels; the creation of centralized digital platforms for consulting the databases of operators and the hospitality businesses; the single management of registrations for public Wi-Fi; actions to update national promotion tools and policies to encourage the digitization of tourist services, as well as tourist materials under public promotion policies (Digital Library, on the model of Europeana), and the management of the related open data.

Another indication provided by Europe has been the push toward the digitization and innovation of museums to promote their knowledge, collections and management sustainability, through the creation of 3D models, augmented reality solutions, gaming experiences and systems of geolocation with the help of personalized museum guides.

To protect citizens and tourists while they plan and gain travel experiences safely, the "Re-open EU" digital platform was established by the Joint Research Center (JRC) in June 2020. This is a constantly updated portal that shows all the innovations made by the Member States and the European Center for Disease Prevention and Control (ECDC) to provide practical and always up-to-date information in relation to borders, transport, tourist services, the pandemic situation of the destinations and health regulations in practice.

Furthermore, in the Communication produced by the European Commission it was indicated that specific economic support is provided for tourism companies to get through the trying circumstance (including, e.g., robots with the function of disinfection and cleaning, crowd management, intelligent booking systems, etc.).

For years, Europe has already been investing in digitalization: in fact, setting up the Digitization Index of the Economy and Society (DESI) of 2020 was a positive step, making headways in all member states and in all the main sectors measured in the index, precisely because digital technologies have become a top priority as a result of the lockdown.

Unfortunately, the pandemic has also shown that Europe, compared to other continents such as Asia and America, lagged behind in digital tourism. The 2020 data showed how much the companies in the sector were underdeveloped in this sphere.

Given the circumstances, the European Commission has also foreseen general supportive measures for companies in the tourism sector. In the above-mentioned Communication, aid was announced from the "digital innovation hub" (DIH), a program which, in the context of Digital Europe in a broad sense, helps to ensure that every company, regardless of size and high-tech ability, has equal digital opportunities. Specifically, within the competitiveness of organizations and the small and medium-sized enterprises (COSME) program, for example, in November 2020, the European Union launched the call for "Boosting the uptake of digitalization, innovation and new technologies in tourism through transnational cooperation and capacity building." This is a European foundation that was designed to strengthen the capacities of small- and medium-sized enterprises in the tourism sector in their transformation toward digitalization, and in the adoption of innovations and new technologies such as data management and AI, promoting transnational cooperation and strengthening skills and competences in the use of new digital technologies. It was created in particular to give life to innovative business models in tourism also through partnerships between the public and private sectors.

On the basis of the indications from Europe, the Spanish Ministry of Industry, Commerce and Tourism worked during the first months of the pandemic to equip the country with new strategic tools based on big data and technological solutions, which were identified as key elements for the relaunch of the tourism sector and to maintain Spain's competitive leadership in the global market (WEF, 2019).

Specifically, Spain launched the Directorio de Soluciones Tecnológicas para los Destinos Turísticos Inteligentes which made it possible to find a series of technological products and services among Spanish digital operators specialized in the tourism sector, thereby creating a balance between supply and demand in the field of technology. The tool has been structured on the basis of the five axes that support the DTI methodology (Destino Turístico Inteligente or Smart Destination) and it is completely free of charge. The catalogue includes almost 100 companies and more than 100 technological solutions ranging from AI, virtual assistants or data analysis systems, to marketing tools, customer relationship management (CRM), video mapping, virtual reality, active listening and web design.

In the summer 2020, Spain also launched the so-called "COVID-tested holidays" on an experimental basis, ensuring the safety of tourist destinations by testing people before departure and upon arrival and offering constant assistance during their stay too. By the time of travel, tourists had to show up at the airport with a negative swab result made at their own expense within 72 hours before departure. During their stay in Spain, all health regulations aimed at guaranteeing maximum protection for travelers had to be strictly followed. Before returning, in the 48 hours prior to departure, the regulation provided that all travelers were given antigenic swabs by designated medical care institutions, at the expense of the tour operator. Thanks to the collaboration with Europe Assistance, travelers were protected by health policies that guaranteed medical assistance and also covered the cancellation of the trip without penalty fee in the event of a positive COVID-19 result. Travelers were therefore reassured by healthy circumstances, in addition to being relieved of most of the bureaucratic and health procedures.

The first pilot scheme for these "COVID-tested holidays" was established on the islands of Fuerteventura and Tenerife by the Alpitour Group, which defined a protocol that later aroused the interest of other destinations such as Greece and Italy.

Still in the technological field, in addition to AI and machine learning, transformation to cloud platforms and investments in cyber security could also promote the recovery of the travel and tourism sectors.

Particular attention also concerns innovations in financial technologies and new payment methods.

In this regard, to meet the growing consumer demand for contactless experiences, Booking.com started a partnership with the startup Futurestay to develop online check-ins for short-term rentals on its platform. In fact, short-term rents seem to have suffered less from the impact of the crisis than hotels, gaining greater favor from travelers who managed to leave during periods of slower spreads of the virus. In addition to Futurestay, Booking.com also collaborates with Hotelbird and Abitari to provide their partners (hosts) with the tools and tailor-made technology they need to offer tourists an effective response to the ongoing evolution of the journey.

Another important strategy some Western countries are focusing on is utilizing blockchains for guaranteeing secure payments and AI for identifying travelers at the airport.

In Italy, for example, the Milan Linate Airport already introduced face boarding in 2020. This biometric technology is based on the facial recognition system (initially active on Milan–Rome flights but progressively extended to other flights). This technology allows passengers to pass security checks and board through an innovative facial recognition system, without having to show their passport and boarding pass at the various checkpoints at the airport. Passengers' personal data are only processed for the purpose of participating in the project. In particular, the facial images are not stored, but are used exclusively to create a biometric model necessary for passing the security checks. The personal data relating to the passports, on the other hand, are kept in an encrypted form for a variable period depending on the consent given by the passenger during registration. Finally, the personal data deriving from the boarding pass are automatically deleted within 24 hours after the flight departure. In any case, travelers can freely choose whether or not to avail themselves of this new system at the moment. In other words, they can still carry out checks-in the traditional way as they used to. Milan Linate Airport also uses other anti-COVID technologies. In particular, the hub uses the new EDS-CB machines with TAC technology that replace X-rays for hand luggage checks. In concrete terms, passengers no longer have to open their baggage to extract liquids, creams, PCs and iPads at security checks. The new technology increases the effectiveness of security checks thanks to the automatic recognition of explosives, and by using high-resolution 3D images, employees are able to check hand luggage quickly and accurately while also shortening the time needed for this. With greater effectiveness in terms of security, waiting time is cut, and there are fewer possibilities of crowding and contact with luggage surfaces.

Furthermore, in 2021 the SEA company carried out a pilot project in collaboration with KME to use surface coatings made out of copper, known for its antiviral and antibacterial properties. All the surfaces in greatest contact with passengers were coated with copper, such as handrails, luggage labels and the bus grab handles and strap stocks. The University of Pisa Institute of Virology has shown that in 60 minutes the viral load of COVID-19 is neutralized by up to 100 percent and in just 10 minutes it is already reduced by 90 percent. Where it is not possible to make the airport contactless, all the necessary anti-infection solutions are in place.

Italy also deserves credit for having designed anticontagion elevator buttons, thanks to an innovative photovoltaic film, so that they can be used without physical contact. The technology, already tested in real contexts and installed in an elevator of the Milan Stock Exchange, was developed by a start-up of the Italian Institute of Technology, Ribes Tech, in the first months of 2021. The idea, originating from the research activity carried out at the IIT Center for Nano Science and Technology in Genoa, is based on the use of special inks and a traditional rotary printer, with which thin and adaptable photovoltaic films are made. These films, integrated with an ad hoc electronic board, can be installed on mechanical buttons, like those of an elevator, making them "smarter" and allowing them to be activated with a gesture of the hand avoiding any physical contact.

The shadow of the hand on the button is interpreted by the photovoltaic film as a variation of light and then is communicated to a recognition system which determines whether the activation is valid or not. The smart button thus provides the electrical signal both when it is mechanically pressed and when it is activated by a hand gesture. The technology, called Daphne Pv, can be applied to an entire elevator push-button panel without the need for significant technical interventions, and after installation it does not require any further certification.

Finally, there is no lack of robotic cleaning systems, home automation and other options that reduce contact with other people. In the United States, for example, "Light Strike" robots created by the American company Xenex have been in circulation since the first months of 2020. These are able to sanitize very large spaces in a very short time, also effectively cleaning furnishings, handrails, handles and other objects or corners that a human cleaner cannot reach easily. These robots are used both in public places such as shopping malls, stations, airports and major hotel chains and in private spaces that require thorough cleaning. The equipment works with high-intensity ultraviolet light, produced by xenon flash lamps, across the entire disinfection spectrum (known as "UV-C"). This energy passes through the cell walls of bacteria, viruses, fungi and spores. A study conducted in Texas at the Biomedical Research Institute (Simmons *et al.*, 2021) found a 99.99 percent reduction in the pathogenic load on complex surfaces as microorganisms' DNA is divided. Light Strike robots operate autonomously and do not need the presence of personnel. Thanks to a particular safety system, they are able to detect every movement in the space in which they are used, instantly blocking their operation to prevent any accidents.

The device has been adopted in a variety of locations, and in full operation in Italy, for example, at the Medici Palace of Ottaviano (near Naples) that currently houses the headquarters of the Vesuvius National Park. In France, however, it is mainly used for sanitizing public hospitals.

Conclusions

COVID-19 is changing the way we live on the planet, but it is also shedding new light on the benefits that technologies and digital innovation can bring to our society in general, and to the recovery of tourist mobility in particular. In this scenario, digital transformation processes must accelerate.

The so-called smart cities, where traditional networks and services function more efficiently through digital technologies and telecommunications for the benefit of these cities' inhabitants and economies are territories where coexisting with COVID-19 is now less painful.

Consequently, the ability to build citizen-friendly cities and implement crisis management effectively add weight to the competitiveness of a country in the contemporary global tourism market.

In the post-COVID era the travel and tourism industry must re-emerge more strongly, more sustainably and more digitally than ever. There is a great opportunity to prioritize initiatives that can accelerate recovery through the digital transformation of the travel and tourism industry. Technology will therefore play a crucial role as a recovery engine and accelerator.

In a smart city that helps people to coexist with COVID-19, the economy (Smart Economy) is based on digital innovation and advanced smart working models. There, people with good digital skills (Smart People) play an increasingly active role in the community in which they live, thanks to the use of new communication tools. The Public Administration (Smart Administration) dialogues directly with citizens and expands the range of digital services. Mobility (Smart Mobility) is flexible, light and ecological. The environment (Smart Environment) is safeguarded by a better use of resources. Living (Smart Living) is supported by easier access to healthcare, education and culture.

Globally new technologies can significantly contribute to counteracting the transmission of the virus. Thanks to applications that monitor people's movements, infections can be traced and their spread slowed down. The processing of the large amount of data from tracking offers important epidemiological indications. Big data analysis can, in fact, provide useful information on how the virus spreads, which situations are most risky or which social categories are most exposed.

In addition, tracking technologies provide a concrete assessment of the emergency in a given territory, allowing people to make targeted choices based on the data collected. The skillful use of tracking systems in China, and South Korea serves as an example for European countries, which lag behind in this field.

Several forecasting studies (EY, 2020; European Commission, 2020b; EIB, 2021) agree that in the next 6–7 years Europe will advance in the smart evolution. This digital transformation will involve hardware, services and software in public and private transport, logistics, multiutility, energy, education, health, security, construction and tourism sectors. The spur to change has been created both by the shock of the pandemic and by the approval of the European Green Deal, reinforced by the financing of the recovery plan that will give rise to the Next Generation EU Plan.

The pandemic has taught us that global collaboration in dealing with health emergencies is central to overcoming crises in many sectors. Networks of cities should work to improve the standardization of protocols for greater sharing of data which, in the event of an epidemic or disaster, lead to a global improvement in the understanding and management of events. The asymmetry of information, on the other hand, only gives a partial view of the situation in urban places.

Restrictions on the movement of people have intrigued interest in domestic tourism. As anticipated, in China, for example, the Chinese New Year was celebrated successfully, with an increase in the number of domestic visitors. Be it domestic or international tourism, cities must also become smart in transport and control systems in an attempt to comply with security regulations. These changes are necessary to disperse people and avoid crowding.

The first measure already widely adopted in this regard was the improved method of booking and purchasing tickets in digital form, with electronic payment. This strategy is useful to avoid queues, avoid the circulation of banknotes and track the movements of passengers and tourists. Countries with underdeveloped online payment systems must work toward updating them quickly. Similarly, also in this scenario, some countries have focused on the analysis of big data and have thus been able to understand more deeply the flows of travel and consequently re-design their urban mobility systems.

In the field of private transport, during the emergency period in Europe many national incentives were allocated for the purchase of bicycles and vehicles for micro electric mobility. These initiatives not only go in the direction of decongesting public transport, but are also designed to reduce the latter's environmental impact. Lighter and more flexible mobility is in fact necessary to lessen urban pollution which, in addition to being a contributing cause of climate change, can worsen the damage caused by COVID-19.

Likewise, the AI systems of cities can boast a wide range of technological products that can help both the management of interpersonal relationships, monitoring that the minimum safety distances are respected, and the early diagnosis of epidemics.

In fact, in many contexts, sensors that minimize the points of contact between strangers are becoming more widespread, as in the case of staff and consumers. Digital tools, such as software, apps and digital menus have proved to be of fundamental importance as antipandemic elements.

Recently, modern people-counters have developed the Intelligent People Counter, which is the updated version of the sensors that were already traditionally used in commercial spaces to monitor traffic in terms of access, the demographic composition of customers (e.g., gender and age) and flows of movement within spaces. The same technology has been enhanced to ensure the safety of workers and visitors. Modern people-counters not only monitor accesses and analyze traffic flows, but also calculate the distance among people. Their simplest use is to control entrances, ensuring that the number of customers never exceeds the critical limit and keeping them at the safety distance by automatically counting entrances and exits.

A people-counting sensor, connected to a screen or to other machines, can help retailers and businesses to automatically regulate inflow. If the sensors are sensitive to infrared, they can also be used as a thermometer for the new customers entering a structure, as per new provisions.

Moreover, the technology also leads beyond these basic functions. Today's people-counters, for example, can tell, through face analysis, whether or not people are wearing masks as required by the security regulation for offices and companies, or even measure the distance among people present in a room, alerting security managers in the event of violations, thus helping to ensure that the measures implemented are working.

The implementation of the latest technology has also been tested to cope with the new and rigorous requirements for sanitizing and cleaning spaces. Robot cleaners can vacuum, scrub floors, clean windows, but also constantly sanitize sensitive surfaces like door handles, elevator buttons and light switches. Bipolar ionization or electrostatic technology both offer greater coverage for efficient decontamination.

In forthcoming years, these higher cleaning standards will be increasingly sought after by local inhabitants, tourists and visitors as a method to reduce the numerous points of contact in shared spaces and to create a more exclusive experience but a more sanitized environment from a social point of view.

In terms of health regulation too, there are thermal cameras or sensors driven by the Internet of Things (IoT). While thermal imaging cameras alone aren't enough for pandemic detection, as is the case of COVID-19, they can offer a number of benefits if used with AI products. The fact that they measure the temperature of suspected COVID-19 cases at airports and other crowded places is proof of their potential in an automated way. Data from various technology products can help enrich health databases, thus providing more accurate information efficiently.

The ongoing debate, about the usefulness and use of smart products and tools, however, centers around two contradictory positions between the need to share useful data for saving life and the risk of damaging the right to privacy.

Some experts suggest that the IoT devices in use should support open protocols but, at the same time, the device vendor should ensure that data integrity and security are respected during their communication and transmission (e.g., Liu *et al.*, 2015; Machado and Medeiros Fröhlich, 2018; Ang, 2020; Zhao *et al.*, 2020).

As anticipated, there is a controversial situation regarding the prospects for the collection and management as well as the development of data systems. The privacy of information and the ways in which it is treated and protected are sensitive issues of great importance. Data access is a challenge for many because information is often considered sensitive for reasons of national security, but, at the same time, it is widely agreed that a virus outbreak poses a dangerous threat to both national and global security.

The advent of COVID-19 has stimulated with varying results a process of increasing normalization of surveillance, which many people in the world were perhaps unwittingly already getting used to. In fact, the role of controllers assumed by some technologies is redefining the value that people place on privacy. The spread of and access to the Internet on a global level and the advent of big data have contributed to an anthropological and social mutation. As we know, people are increasingly connected through the web. Their desires, their needs, but also their behaviors are becoming increasingly exposed and in the public domain. They are disclosed through posts on social networks or captured through some algorithms when Internet users accept certain navigation conditions. For these reasons, the issue of protecting private information and people's freedoms is a critical issue. We need to find the most suitable way to establish an equilibrium between the individual need for data protection and the collective interest.

Therefore, since this data could spread across different countries, it would be advisable to identify standardized and universally agreed protocols, at least in the event of an emergency. Despite all the advantages offered by technology in terms of strengthening measures designed to contain the COVID-19 pandemic, it is questionable whether government decisions that ignore or overshadow the right to privacy can be justified in order to provide a coordinated and timely response to health emergencies. The risk many fear is that such measures could become permanent. Consequently, the trade-off just described must be evaluated very carefully.

Against this backdrop, emerging technologies, such as blockchain or quantum cryptography, can help and integrate with data-collection technologies, potentially providing more data from both the medical sector and smart city operators, while ensuring greater privacy and security. Such a system, therefore, would provide the relevant information to help decision makers formulate regulations for the maximum protection of both clonal populations and travelers.

Bibliography

Akiyama, I (2020) 'Basic and recent applied technologies of ultrasound in the field of clinical diagnosis', *J-Stage*, 41, 6: 845–850.

Amadeus (2020) *Destination X: Where to Next—What Leisure Travelers Want*, Madrid: Amadeus IT Group SA.

Amann, J, Sleigh, J and Vajena, E (2020) 'Digital contact-tracing during the covid-19 pandemic: An analysis of newspaper coverage in Germany, Austria, and Switzerland', *PLoS ONE*, 2, 3.

Ang, Y (2020) 'When COVID-19 meets centralized, personalized power', *Nature Human Behaviour*, 4: 445–447.

Anguera-Torrell, O, Aznar-Alarcón, J P and Vives-Perez, J (2020) 'COVID-19: Hotel industry response to the pandemic evolution and to the public sector economic measures', *Tourism Recreation Research*, 1: 1–10.

Anichiti, A, Dragolea, L L, Tacu Hârşan, G D, Haller, A P and Butnaru, G I (2021) 'Aspects regarding safety and security in hotels: Romanian experience', *Information*, 12, 1: 44.

Ashikul, H, Farzana, A, Mohammad, W and Ishtiaque, A (2020) 'The effect of coronavirus (COVID-19) in the tourism industry in China', *Asia Journal of Multidisciplinary Studies*, 3, 1.

Assaf, A and Scuderi, R (2020) 'COVID-19 and the recovery of the tourism industry', *Tourism Economics*, 26, 5: 731–733.

Babu, G and Justin, P (2020) *Digital Transformation in Business and Society. Theory and Cases*, Cham: Palgrave Macmillan.

Bamforth, C (2006) 'The capital delights of Nara', *The Japan Times*, 05/26/2006.

Bishop, D (2005) 'Lessons from SARS: Why the WHO must provide greater economic incentives for countries to comply with international health regulations', *Georgetown Journal of International Law*, 36, 4: 1173–1226.

Booking (2020) *Global Research Report—Future of Travel*, Amsterdam: Booking.com.

Boulos, M N, Brewer, A C, Karimkhani, C, Buller, D B and Dellavalle, R P (2014) 'Mobile medical and health apps: State of the art, concerns, regulatory control and certification', *Online Journal of Public Health Informatics*, 5, 3: 229.

Calvo, R A, Deterding, S and Ryan, R M (2020) 'Health surveillance during Covid-19 pandemic', British Medical Journal, 369: m1373.

Castells, M (1996) *The Rise of the Network Society*, Malden: Blackwell Publishers.

Castells, M (2000) *End of Millennium*, Malden: Blackwell Publishers.

Castells, M (2005) 'Global governance and global politics', *Political Science & Politics*, 38, 1: 9–16.

CBI (2019) *Tech Tracker 2019: The Must-Know Technology and Innovation Trends*, New York: CBI.

Chandra, S (2013) 'Deaths associated with influenza pandemic of 1918–19', *Japan. Emerging Infectious Diseases*, 19, 4: 616–622.

Chan, I C C, Ma, J, Ye, H and Law, R (2021) 'A comparison of hotel guest experience before and during pandemic: Evidence from online reviews', *Information and Communication Technologies in Tourism* 2: 549–556.

China Tourism Academy (2021) *China Leisure Development Annual Report*, Beijing: China Tourism Academy.

Cooper, I, Mondal, A and Antonopoulos, C G (2020) 'Dynamic tracking with model-based forecasting for the spread of the COVID-19 pandemic', *Chaos, Solitons & Fractals*, 139: 110298.

Crawford, K, Miltner, K and Gray, M L (2019) 'Critiquing big data: Politics, ethics, epistemology', *International Journal of Communication*, 8: 1663–1672.

deLisle, J (2004) 'Atypical pneumonia and ambivalent law and politics: SARS and the response to SARS in China', *Temple Law Review*, 77: 193–245.

Deng, J. (2020) 'Taiwan Stopped Covid-19's Spread, but Can't Talk About It at WHO Meeting', *The Wall Street Journal*, 11/12/2020.

Dušek, R, Sagapova, N and Kliestik, T (2021) 'Effect of the COVID-19 global pandemic on tourists' preferences and marketing mix of accommodation facilities. Case study from Czech Republic', *SHS Web of Conferences* 92: 1009.

ECDC (2021) *Situation Updates on COVID-19*, Solna: European Centre for Disease Prevention and Control.

EIB (2021) *Building a Smart and Green Europe in the COVID-19 Era*, Luxemburg: European Investment Bank.

European Commission (2020a) *Tourism and Transport in 2020 and Beyond*, Brussels: European Commission.

European Commission (2020b) *The Digital Economy and Society Index (DESI)*, Brussels: European Commission.

EY (2020) *Smart City Report*, London: Ernst&Young.

Farris, W W (1985) *Population, Disease, and Land in Early Japan, 645–900*, Cambridge: Harvard University Asia Center.

Fennell, D A (2021) 'Technology and the sustainable tourist in the new age of disruption', *Journal of Sustainable Tourism*, 29, 5: 767–773.

Gallego, I and Font, X (2021) 'Changes in air passenger demand as a result of the COVID-19 crisis: Using big data to inform tourism policy', *Journal of Sustainable Tourism*, 29, 9: 1470–1489.

Gargi, K and Maitri, M (2015) 'Gen z—Children of digital revolution transforming social landscape', *American International Journal of Research in Humanities, Arts and Social Sciences*, 15, 369: 206–208.

Gasser, U and Palfrey, J (2008) *Born Digital: Understanding the First Generation of Digital Natives*, New York: Basic Books.

Gaudin, S. (2014) 'First Robot, Networked Tablets Head to West Africa to Fight Ebola', *Computer World*, 11/26/2014.

Golten Jutel, A (2011) *Putting a Name to It: Diagnosis in Contemporary Society*, Maryland: Hopkins University Press.

Gössling, S (2021) 'Technology, ICT and tourism: From big data to the big picture', *Journal of Sustainable Tourism*, 29, 5: 849–858.

Grundner, L and Neuhofer, B (2021) 'The bright and dark sides of artificial intelligence: A future perspective on tourist destination experiences', *Journal of Destination Marketing & Management* 19: 100511.

Hall, C M, Scott, D and Gössling, S (2020) 'Pandemics, transformations and tourism: Be careful what you wish for', *Tourism Geographies*, 2: 1–22.

Han, B C (2017) *Psychopolitics*, London: Verso Books.

Hendl, T, Chung, R and Wild, V (2020) 'Pandemic surveillance and racialized subpopulations: Mitigating vulnerabilities in COVID-19 apps', *Journal of Bioethical Inquiry*, 17, 4: 829–834.

Hopkins, D R (2002) *The Greatest Killer: Smallpox in History*, Chicago: University of Chicago Press.

Horak, S and Weber, S (2000) 'Youth tourism in Europe: Problems and prospects', *Tourism Recreation Research*, 25: 37–44.

Huang, Y, Sun, M and Sui, Y (2020) 'How digital contact tracing slowed covid-19 in East Asia', *Harvard Business Review*, 4.

Ienca, M and Vayena, E (2020) 'On the responsible use of digital data to tackle the COVID-19 pandemic', *Nature Medicine*, 26, 4: 463–464.

ILCJ (2017) *Population Aging and Aged Society: Population Aging and Life Expectancy*, Tokyo: International Longevity Center Japan.

Ivanova, M, Krasimirov, I and Stanislav, I (2021) 'Travel behaviour after the pandemic: The case of Bulgaria', *Anatolia*, 32, 1: 1–11.

Jaipuria, S, Parida, R and Ray, P (2021) 'The impact of COVID-19 on tourism sector in India', *Tourism Recreation Research*, 46, 2: 245–260.

Jannetta, A B (2014) *Epidemics and Mortality in Early Modern Japan*, Princeton: Princeton University Press.

Japan Ministry of Health (2021) *Handbook of Health and Welfare Statistics*, Tokyo: Japan Ministry of Health.

Kamal, M M (2020) 'The triple-edged sword of COVID-19: Understanding the use of digital technologies and the impact of productive, disruptive, and destructive nature of the pandemic', *Information Systems Management*, 37, 4: 310–317.

Kaushal, V and Srivastava, S (2021) 'Hospitality and tourism industry amid COVID-19 pandemic: Perspectives on challenges and learnings from India', *International Journal of Hospitality Management*, 92: 102707.

Kawana, A, Naka, G, Fujikura, Y, Kato, Y, Mizuno, Y, Kondo, T and Kudo, K (2007) 'Spanish Influenza in Japanese armed forces, 1918–1920', *Emerging Infectious Diseases*, 13, 4: 590–593.

Klenk, M and Duijf, H (2020) 'Ethics of digital contact tracing and COVID-19: Who is (not) free to go?', *Ethics and Information Technology*, 8: 1–9.

Knobler, S, Mahmoud, A and Lemon, S (eds.) (2004) *Learning from SARS: Preparing for the Next Disease Outbreak: Workshop Summary*, Washington: National Academies Press.

Kono, Y, Shimizu, E, Matsunaga, F, Egami, Y, Yoneda, K, Sakamoto, K, Mubita, M, Sunkutu, V, Wakamatsu, K, Terashima, M and Fujita, N (2021) 'Enhancing the use of computed tomography and cardiac catheterization angiography in Zambia: A project report on a global extension of medical technology in Japan', *Global Health & Medicine*, 3, 1: 52–55.

Kreps, G L and Neuhauser, L (2010) 'New directions in eHealth communication: Opportunities and challenges', *Patient Education and Counseling*, 78, 3: 329–336.

Lai, C C, Shih, T P, Ko, W C, Tang, H J and Hsueh, P R (2020) 'Severe acute respiratory syndrome coronavirus 2 (SARS-CoV-2) and coronavirus disease-2019 (COVID-19): The epidemic and the challenges', *International Journal of Antimicrobial Agents*, 55, 3.

Lee, D and Lee, J (2020) 'Testing on the move: South Korea's rapid response to the COVID-19 pandemic', *Transportation Research Interdisciplinary Perspectives*, 5: 100–111.

Lee, D, Heo, D and Seo, Y (2020) 'COVID-19 in South Korea: Lessons for developing countries', *World Development*, 135, 105057.

Lee, M, Lee, H, Kim, Y, Kim, J, Cho, M, Jang, J and Jang, H (2018) 'Mobile App-based health promotion programs: A systematic review of the literature', *International Journal of Environmental Research and Public Health*, 15, 12: 28–38.

Liang, F (2020) 'COVID-19 and health code: How digital platforms tackle the pandemic in China', *Social Media & Society*, 6, 3.

Lin, C, Braund, W E, Auerbach, J, Chou, J H, Teng, J H, Tu, P and Mullen, J (2020) 'Policy decisions and use of information technology to fight COVID-19, Taiwan', *Emerging Infectious Diseases*, 26, 7: 1506–1512.

Liu, C, Yang, C, Zhang, X and Chen, J (2015) 'External integrity verification for outsourced big data in cloud and IoT: A big picture', *Future Generation Computer Systems*, 49: 58–67.

Lupton, D (2015) *Digital Sociology*, London: Routledge.

Lyon, D (1988) *The Information Society: Issues and Illusions*, Cambridge: University of Cambridge Press.

Machado, C and Medeiros Fröhlich, A A (2018) 'IoT data integrity verification for cyber-physical systems using blockchain', in R. Bilof (ed.) *2018 IEEE 21st International Symposium on Real-Time Distributed Computing (ISORC)*, Los Alamitos: IEEE Computer Society.

Martinez-Martin, N, Wieten, S, Magnus, D and Cho, M K (2020) 'Digital contact tracing, privacy, and public health', *Hastings Center Report*, 50, 3: 43–46.

Masuda, Y (1981) *The Information Society as Postindustrial Society*, Bethesda, MD: World Future Society.

McKinsey (2020) *How COVID-19 Has Pushed Companies Over the Technology Tipping Point and Transformed Business Forever*, New York: McKinsey & Company.

Minai, M S, Raza, S and Segaf, S (2021) 'Post COVID-19: Strategic digital entrepreneurship in Malaysia', in B S Sergi and A R Jaaffar (eds.) *Modeling Economic Growth in Contemporary Malaysia*, Bingley: Emerald Publishing Limited.

Naumov, N, Varadzhakova, D and Naydenov, A (2021) 'Sanitation and hygiene as factors for choosing a place to stay: Perceptions of the Bulgarian tourists', *Anatolia*, 32, 1: 144–147.

OECD (2020) *Health at a Glance: Asia/Pacific 2020. Measuring Progress Towards Universal Health Coverage*, Geneva: World Health Organization.

Park, S, Choi, G J and Ko, H (2020) 'Information technology-based tracing strategy in response to COVID-19 in South Korea. Privacy controversies', *Journal of the American Medical Association*, 323, 21: 2129–2130.

Peluso, A M and Pichierri, M (2020) 'Effects of socio-demographics, sense of control, and uncertainty avoidability on post-COVID-19 vacation intention', *Current Issues in Tourism*, 1: 1–13.

Pillmayer, M, Scherle, N and Volchek, K (2021) 'Destination management in times of crisis—Potentials of open innovation approach in the context of COVID-19?', *Information and Communication Technologies in Tourism*, 1: 517–529.

Resnyansky, L (2019) 'Conceptual frameworks for social and cultural big data analytics: Answering the epistemological challenge', *Big Data & Society*, 6, 1.

Rowe, F (2020) 'Contact tracing apps and values dilemmas: A privacy paradox in a neo-liberal world', *International Journal of Information Management*, 55, 102178.

Roy, G and Sharma, S (2021) 'Analyzing one-day tour trends during COVID 19 disruption. Applying push and pull theory and text mining approach', *Tourism Recreation Research*, 46, 2: 288–303.

Sharon, T (2020) 'Blind-sided by privacy? Digital contact tracing, the Apple/Google API and big Tech's newfound role as global health policy makers', *Ethics and Information Technology*, 18: 1–13.

Simmons, S, Carrion, R, Alfson, K, Staples, H, Jinadatha, C, Jarvis, W and Stibich, M *et al.* (2021) 'Deactivation of SARS-CoV-2 with pulsed-xenon ultraviolet light: Implications for environmental COVID-19 control', *Infection Control & Hospital Epidemiology*, 42, 2: 127–130.

Singh, S (2020) 'Quixotic' tourism? Safety, ease, and heritage in post-COVID world tourism', *Journal of Heritage Tourism*, 8: 1–6.

Skinner, H, Sarpong, D and White, G R (2018) 'Meeting the needs of the millennials and generation z: Gamification in tourism through geocaching', *Journal of Tourism Futures*, 4, 1: 93–104.

Steinbrook, R (2020) 'Contact tracing, testing, and control of COVID-19. Learning from Taiwan', *Journal of the American Medical Association*, 180, 9: 1163–1164.

Summers, J, Cheng, H J, Lin, H H, Barnard, L T, Kvalsvig, A, Wilson, N and Baker, M G (2020) 'Potential lessons from the Taiwan and New Zealand health responses to the COVID-19 pandemic', *The Lancet Regional Health—Western Pacific*, 4: 100044.

Sun, H, Qiu, Y, Yan, H, Huang, Y, Zhu, Y, Gu, J and Chen, S (2021) 'Tracking reproductivity of COVID-19 epidemic in China with varying coefficient SIR models', *Journal of Data Science*, 18, 3: 455–472.

Tan, S (2010) 'Authoritative Master Kong in an authoritarian age', *Dao*, 9, 2: 137–149.

Tom-Aba, D, Olaleye, A, Olayinka, A T, Nguku, P, Waziri, N and Adewuyi, P *et al.* (2015) 'Innovative technological approach to ebola virus disease outbreak response in Nigeria using the open data kit and form hub technology', *PLoS ONE*, 10, 6: e0131000.

Travel Tech (2021) *Future Travel Enthusiasm in the Age of COVID-19*, Arlington: Travel Technology Association.

Veen, W and Vrakking, B (2006) *Homo Zappiens: Growing Up in a Digital Age*, London: Bloomsbury Publishing.

Walrave, M, Waeterloos, C and Ponnet, K (2020) 'Adoption of a contact tracing app for containing COVID-19: A health belief model approach', *JMIR Public Health Surveil*, 6, 3: e20572.

Wang, C J, Ng, C Y and Brook, R H (2020) 'Response to COVID-19 in Taiwan: Big data analytics, new technology, and proactive testing', *Journal of the American Medical Association*, 323, 14: 1341–1342.

Watanabe, T and Omori, Y (2020) *Online Consumption During and After the COVID-19 Pandemic: Evidence from Japan*, Tokyo: CREPE.

WEF (2019) *Travel & Tourism Competitiveness Report 2019*, Genève: World Economic Forum.

West, D M (2015) *Using Mobile Technology to Improve Maternal Health and Fight Ebola: A Case Study of Mobile Innovation in Nigeria*, Washington: Center for Technology Center for Technology Innovation at Brookings.

WHO (2020a) *Report of the WHO-China Joint Mission on Coronavirus Disease*, Geneva: World Health Organization.

WHO (2020b) *Updated WHO Recommendations for International Traffic in Relation to COVID-19 Outbreak*, Geneva: World Health Organization.

Wu, Y C, Chen, C S and Chan, Y J (2020) 'The outbreak of COVID-19: An overview', *Journal of the Chinese Medical Association*, 83, 3: 217–220.

Zhao, Q, Chen, S, Liu, Z, Baker, T and Zhang, Y (2020) 'Blockchain-based privacy-preserving remote data integrity checking scheme for IoT information systems', *Information Processing & Management*, 57, 6.

3 How to choose safe destinations

From "communitycation" to new digital analysis tools

Introduction

Prior to the advent of postindustrial society, the boundary between leisure and work time had for a long period been distinctive, and profoundly affected travelers' lives. From the late 1970s to the early 1980s, things gradually began to change. We must not forget that tourist experiences in industrial society were marked by the free time left by working activity. In these contexts, paid holidays represented the only time dedicated to traveling, and abandoning the daily routine meant having the opportunity to relax, rest and refresh, in anticipation of the recommencement of the working week. When the tertiary sector began to reign over the others, with a growth in the production of services, the so-called postindustrial society took shape (e.g., Bell, 1973; Featherstone, 1991). The distinctive features of postindustrialization were, among others, the end of the wage society, which drew the distinction between productive and nonproductive time, and the spread of values and cultures centering on free time and the prevalence of narcissistic attributes (e.g., Kumar, 1995; Holt, 1997; Campbell, 2005). At this point, tourist experiences found their place even outside the main periods traditionally dedicated to holidays (e.g., in occasion of cultural, religious or national celebrations), also following the introduction of flexibility in the world of work (e.g., De Masi, 1985; Dann, 2002).

Thus, in the postmodern era, the stereotypical holiday and the tourist environment, previously proposed solely as a moment of playful heterodirection or as a mark of social status, have given way to holidays and tourist experiences with profoundly different characteristics. In fact, postmodern tourists immediately appeared more demanding, informed, experienced and price-sensitive than previously (e.g., Lanfant, Allcock and Bruner, 1995; Nuryanti, 1996; Larsen, Urry and Axhausen, 2016). For them, tourism had become an ordinary experience, a daily activity undertaken for very disparate reasons, depending on the meaning that individual social actors attributed to the journey they intended to undertake. In an increasingly cosmopolitan and globalized society, the "tourist gaze" (Urry, 1990) began to turn to all those destinations that in the eyes of individual travelers seemed

DOI: 10.4324/9781003195177-4

to offer an enriching, out-of-the-ordinary experience. The tourist area has thus expanded beyond the specialized areas of historic cities, mountain areas and bathing areas. It has become increasingly evident that tourist attractions can no longer be limited solely to places, monuments or events. Travelers from all over the world have gradually begun to divert their attention to local people, sociocultural conditions and local customs and traditions. From a sociological perspective, we can argue that the complexity of postmodern society has meant that tourist experiences have multiplied (e.g., Urry, 1995; Ryan and Glendon, 1998; Salazar, 2006). The image of the tourist is no longer that of an inactive customer of the tourism industry, but rather it has taken on the features of a curious explorer, in possession of tools and resources, being more circumspect, aware and prepared. Contemporary tourists have been able to rely not only on technologies, but above all on the skills to handle them, to gain access to the tourist product of their interest without having to turn to any third parties (e.g., Stamboulis and Skayannis, 2003; Rosenbloom, 2007; Kaewkitipong, 2010).

It has emerged quite clearly that in the contemporary world tourism has become a social and cultural experience promoting both paths toward socialization and the construction of an identity, where subjects experience being in a free and fluid state (e.g., Palmer, 1999; Berman and Bruckman, 2001; McIntosh, Hinch and Ingram, 2002). In other words, in postmodernity, travel choices are no longer conditioned exclusively by the tourism industry. Social actors have gradually gained a greater awareness of themselves and their needs. Consequently, the tourist trip has been transformed into a possible experience capable of satisfying tourists' needs, also shaping their identities. Within this framework, tourists' personalities appear highly multifaceted: their actions are driven by a hedonistic, ludic component, but also by the desire to enrich their own profile, through direct or indirect experiences and knowledge. In this sense, tourists can be defined as autonomous, self-determined, active and passionate subjects (e.g., Choen, 2004; Gmelch, 2004; Trauer and Ryan, 2005; Urry and Larsen, 2011). It is safe to argue that the boundary between tourists and travelers has been blurred, since it is pointless to make any attempt to fit the concept of a tourist into a one-dimensional, static and definitive representation.

Under the pressure of technological advancement, the extension of spaces and the acceleration of mobility, the tourist market has experimented with new forms of the organization, planning and preparation of the tourist experience, tailoring it to the tastes and expectations of a tourist's search for experiences in which identities and communities are produced.

The advent of COVID-19 has clearly represented a turning point in experiencing and understanding tourism. Today tourists have to find a compromise between their own mobility needs and deeply transformed international circumstances. As mentioned in previous chapters, tourists have to confront not only the severe restrictions imposed on mobility, but also substantial differences in terms of the levels of spread of the virus and the differing local health rules put in place to combat the pandemic among various territories.

As a result of this, at least two new decisive factors affect the choice of a tourist destination. First, of course, the perception of safety; secondly, the central role of the possibility of creating a worry-free tourist experience. On the first point, the new rule that travelers are following (with increasing frequency) is to choose where to go by carefully evaluating the level of travel risk. Whether they want to make a very organized international or intercontinental trip, or to be a backpacker in nearby territories, they must still be aware of the health conditions of the destination they choose to visit. Therefore, some destinations will lose appeal once considered too dangerous or because the local health system is not considered sufficiently adequate in case of emergency or necessity.

Similarly, countries around the world show differences in precautionary measures, having regional variations even within the same areas. These can range from total lockdown to the closure of commercial activities, from curfews to a ban on events and initiatives. Travelers today are aware of the possibility of a restricted experience while traveling to their destination due to their inability to experience what was possible before the outbreak of COVID-19. For example, the closure of specialty restaurants makes it impossible for tourists to taste typical local cuisine, or the closure of museums and cultural centers deprives tourists of the opportunity to learn more about their destination. Furthermore, complying with the local health protocols of the chosen destination and the departure point, travelers may be obliged to receive mandatory health tests, or, forced self-isolation, quarantine or additional tests, before departure. In Europe, for example, the European Union (EU) has divided the regions into green, orange, red and dark red zones, based on the respective COVID-19 infection rates. The dark red category was introduced later to indicate the areas where the virus circulates at very high levels, also due to the level more COVID-19 variants (Council of the European Union, 2021).

In early 2021, the European Commission also recommended stricter measures for international travelers entering the EU from non-EU countries. International travelers would be required to complete and submit a health declaration statement, used by Member States to reconstruct the network of contacts that people had during their mobility experience.

Other "green areas," in which the circulation of COVID-19 is almost vanishing, have also been identified in the world. Among these areas people can move without particular limits.

A first example in this sense is represented by New Zealand. At the start of the pandemic, the New Zealand government followed an approach similar to that of many other countries: it tried to fight the spread of the disease by using existing pandemic plans originally designed for influenza, thus aiming keep the workload for the health system within tolerable levels. After a few weeks it became evident that the pandemic plan measures implemented by New Zealand were proven to be inadequate, so the country started strict lockdowns that helped reduce the number of new positive cases and therefore the circulation of COVID-19.

In the spring of 2020, European countries followed a similar approach to lockdown, but what differentiates them from New Zealand is the fact that the latter maintained a high level of control even in the following few months. There, the limitations along the borders lasted for months from the start of the lockdown and the contact tracing activities did not cease with the help of technologies. In several cases, the identification of a handful of new cases was enough to impose new strict lockdowns to avoid new outbreaks and the circulation of COVID-19, which would render the efforts made in the previous months futile. A similar approach was followed in Australia too. For example, the state of Victoria initiated a lockdown of about 4 months in July 2020 and later adopted severe new restrictions followed by the detection of only 13 new positive cases in 2021.

Clearly, New Zealand and Australia have managed to contain the pandemic also thanks to their special geographic structure as island countries, that is, not bordering other countries. The possibility of controlling the inflows and outflows of people more rigidly than elsewhere has had a significant impact, and a similar phenomenon has been observed in other countries with the same characteristics.

Iceland, for example, managed to reduce new infections and keep the pandemic better under control even while several European countries were struggling with the "second wave." Also, in this case the main advantage of Iceland being an island derives from the possibility of controlling the flow of people better, not to mention the country's different and lower population density compared to many European States.

In this uncertain, differentiated and constantly changing scenario, it is clear that the cognitive needs of travelers have multiplied. They need to know not only where they can go, mediating between their mobility wishes and the restrictions imposed at national and supranational levels, but they also feel the need to be constantly updated on what they can possibly do in the individual territories. To meet these diverse requirements and support tourists from all over the world, various initiatives have emerged in recent times. This chapter introduces and analyzes the main solutions that allow people to choose a destination in a pandemic-sweeping world.

3.1 Beyond the official sources: The tourist communitycation

Before presenting some platforms that present and compare different destinations based on a series of useful indicators that orient travelers toward one place or another, a small parenthesis must be given to online word-of-mouth.

A large amount of empirical data illustrate different aspects of the pandemic and its harmful effects are constantly spreading through the Internet (e.g., Brainard, 2020; Fetzer, Witte and Hensel, 2020; Rzymski *et al.*, 2020). On the one hand, this lets people receive informal information on the impact of the pandemic, but on the other, it risks creating situations of overlapping and confusion, especially among people without the necessary skills or tools needed to be able to interpret the information they come across online correctly.

If this bottom-up production of information is associated with a lack of confidence in institutions, which have not always been considered capable of managing the pandemic in the best way (e.g., Bargain and Aminjonov, 2020; Kooistra and van Rooij, 2020; Seytre, 2020; Parmet and Paul, 2020; Yamin, Habibi and Rights, 2020; de Campos-Rudinsky and Undurraga, 2021), then it would be clear to see why in the current scenario a central role has been occupied by what I have elsewhere (Monaco, 2019) defined as "tourist communitycation."

This composite word is derived from a combination of two terms: "community" and "communication" and refers to the communication that takes place within online virtual communities, mainly represented today by social networks. As recent studies have highlighted (e.g., Arpaci *et al.*, 2020; Boon-Itt and Skunkan, 2020; Bruns, Harrington and Hurcombe, 2020; Jimenez-Sotomayor, Gomez-Moreno and Soto-Perez-de-Celis, 2020; Lai *et al.*, 2020; Rovetta and Bhagavathula, 2020) since the onset of the pandemic, social media quickly acquired a central role in the creation, dissemination and consumption of information.

In the midst of the pandemic, the communication activated before the organization of a physical trip is aimed at collecting data and information on the destinations on the precautionary measures in place and the incidence of the infection among citizens (e.g., Abd-Alrazaq *et al.*, 2020; Abdullah *et al.*, 2020; Dandapat *et al.*, 2020; Corbisiero and Monaco, 2021). In this scenario, online word-of-mouth involves both tourists who are evaluating different alternatives, and local inhabitants. They are considered privileged witnesses who can dispel all doubts or answer specific questions, such as opening hours of the local services and businesses or rules on quarantine and curfew.

The tourist communitycation is also very useful in purchasing decision, during which people decide to buy one or more service and evaluate across different channels. The main concerns include buying travel tickets, choosing and booking accommodation facilities, and more generally tourist packages. In this case, tourists ask people who have already made the same choice for information on the possibility of reimbursement or change of date in case of cancellation of the trip or tightening of restrictions. Memoirs, reports, reviews, opinions, stories and anecdotes, together with the information generated by tour operators and institutions, all constitute the description of travel destinations.

One of the fundamental effects that can be traced in this dynamic is represented by the "fiduciary" social exchange among subjects, since the communitycation that takes place in these virtual spaces is not hierarchical, yet participatory.

From this, analysis, data, news and information shown on the sites and shared through online chatting and social networks, satisfy contemporary tourists' needs to broaden their horizons.

3.2 Updated maps and websites

Countries and territories around the world have created specific web pages on their institutional websites offering citizens and potential tourists the opportunity to monitor the pandemic situation in real time, from the trend of the contagion curve to the mortality rate, and the number of deaths.

Globally, an initiative toward this goal is the "WHO Coronavirus Disease (COVID-19) Dashboard" (COVID19.who.int), edited by the World Health Organization. The site is constantly updated and contains information on the number of vaccines administered, confirmed cases and new positives all around the world. On it web users can see data represented on the world map and understand the situation in each country by shade of color, territory or area, indicating in the search engine the specific country's name. In addition, data can be requested by day, week and month, and displayed in forms of graphs or tables (see Fig. 3.1).

There are also sites with continental dashboards that contain the latest government announcements regarding containment measures and restrictions for travelers.

For example, in the United States, the tracking activity is conducted by the Center for Disease Control and Prevention, which is one of the main operational components of the Department of Health and Human Services.

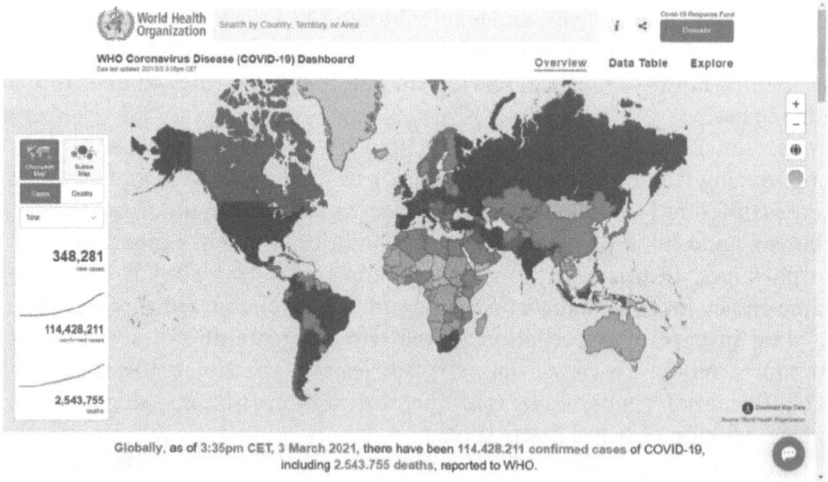

Figure 3.1 WHO Coronavirus Disease (COVID-19) Dashboard homepage.

Source: https://covid19.who.int (accessed March 04, 2021).

The Center works tirelessly to protect American citizens at home and abroad from threats of health, safety and security by supporting the community through a variety of activities.

As a national agency for health protection, the American Center for Disease Control and Prevention has cooperated with the Georgia Tech Research Institute to establish the so-called "COVID Data Tracker" (www.covid.cdc.gov) to disseminate relevant information to help decision-making by the public and travelers, with the purpose of reducing risks and keeping track of new transmissions (Citroner, 2020). The web portal contains a series of maps and charts tracking the cases, deaths and trends of COVID-19 in the United States, and the information renews at 8 p.m. on a daily basis (see Fig. 3.2).

The site also contains information on travel restrictions and precautionary measures that travelers must comply with in each US state. The interactive map combines documented county-level clinical cases and other population surveys. Using a binomial probability model, it also estimates the circulation rates of COVID-19 day by day (Chande *et al.*, 2020). Recognizing the difficulty in tracking all COVID-19 cases, the tool calculates the risk with the assumption that the current number of infections is up to ten times higher than what official reports indicate. Indeed, models and tools like this can provide perspective, but are limited as they may not be accurate and often cannot account for every variable. A critical issue may be, for example, that the number of cases would be probably underestimated due to test deficiencies, the existence of asymptomatic people or communication delays (Yi *et al.*, 2007).

Figure 3.2 COVID Data Tracker's homepage.

Source: https://covid.cdc.gov/covid-data-tracker (accessed: March 04, 2021).

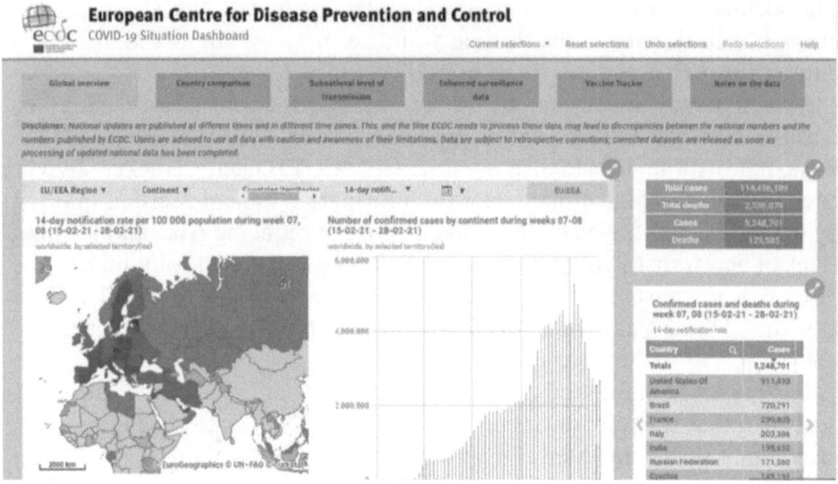

Figure 3.3 COVID-19 Situation Dashboard's homepage.

Source: https://www.ecdc.europa.eu/en/covid-19/situation-updates (accessed: March 04, 2021).

Similarly, on the European side, the European Center for Disease Prevention and Control has prepared the "COVID-19 Situation Dashboard" (qap.ecdc.europa.eu), which is a series of maps published every Thursday (see Fig. 3.3).

In response to the COVID-19 pandemic, this dashboard supports the EU Council with a coordinated approach, focusing on the policy of restricting-free movement adopted by the EU Member States on October 13, 2020 (Council of the European Union, 2020). Once a week the EU Member States report pandemic-related data to the European Surveillance System database (TESSy) and then the site makes the maps with different colors showing the epidemic intensity of each country:

- Green, if the 14-day notification rate is lower than 25 cases per 100,000 inhabitants and the test positivity rate remains below 4 percent.
- Dark Gray, either if the 14-day notification rate is lower than 50 cases per 100,000 inhabitants and the test positivity rate is 4 percent or higher, or if the 14-day notification rate is between 25 and 150 cases per 100,000 inhabitants and the test positivity rate is below 4 percent.
- Black, either if the 14-day cumulative COVID-19 case notification rate ranges from 50 to 150 cases per 100,000 inhabitants and the test positivity rate for COVID-19 infection is 4 percent or more, or if the 14-day cumulative COVID-19 case notification rate is more than 150 but less than 500 cases per 100,000 inhabitants.
- Dark Red, if the 14-day cumulative COVID-19 case notification rate is 500 cases per 100,000 inhabitants or higher.

- Gray, if there is insufficient information or a testing rate lower than 300 cases per 100,000 inhabitants.

Something similar also exists in China: AliHealth, a subgroup of the e-commerce giant Alibaba, has created a mini-program attached the Alipay app, showing various data concerning the COVID-19 situation both home and abroad as well as incorporating various other functions (see Fig. 3.4). More specifically, the mini-program presents very detailed information about the situation of COVID-19 in China: the first set of data that comes to attention shows the user's current local situation with the aid of GPS with categorized data of new confirmed cases, new asymptomatic cases, current confirmed cases, and the total number of patients who have recovered. With just a click, a city map will pop up and give a holistic view of the details in each district of the zone, showing colors of different shades in accordance with the different intensity of the infection situation. In addition, there is also a chart showing data from specific provinces and cities with the subdivision of imported cases and local cases. All the data is a reorganization of the public information from the Chinese National Health Commission, the health commissions of every province, city and district, and other official sources such as the WHO. Apart from these useful data, the mini-program

Figure 3.4 AliHealth mini-program on Alipay.

Source: Alipay app (accessed: March 21, 2021).

has a navigation bar that displays useful functions related to people's lives: appointment and results of nucleic acid testing, local regulations regarding work resumption, risky areas, latest news and regulations at the national, provincial and city level, as well as the trajectory of the virus carrier after the epidemiological survey. An especially noteworthy function is the refutation of rumor, which is a section dedicated to disprove online rumors about the coronavirus and offer relevant scientific information, enabling people to combat the spread of the disease with scientific awareness. Since the vaccination campaign has started across China, the mini-program has also been updated with a function that helps user to know the status quo of research and development of different vaccines against COVID-19.

As the most popular instant messaging software, WeChat also has a similar anti-COVID mini-program, which is found in the subsection of the city service of daily life. Similar to that of AliHealth, this mini-program contains information taken from the Chinese National Health Commission and is updated on a daily basis. The mini-program gives a general view of the virus situation both at home and abroad and also the trend and intensity of each city or country. Apart from these data and functions, similar to that of AliHealth, it has a special function used for epidemiological surveys: the user can enter the district name, flight number or train number to check to see if any virus carrier has been in the same place along the journey.

If people intend to travel or would like to find more detailed information on the destinations they are about to visit, they can also cast their doubts away by visiting the TripsGuard website (www.tripsguard.com). This is an interactive map that allows people to choose among possible destinations and find out what measures are in force. The portal has been developed by Avian, a technology company involved the sale of airline tickets. It is updated often, but not in real time. TripsGuard collects information from multiple sources and aggregates them for each country. Being always a "work in progress," it exploits collaborative functions to allow its users to report inaccuracies or outdated indications.

A similar proposal is the "COVID-19 Travel Regulations map" (www.iatatravelcentre.com/world.php) powered by Timatic and designed by the International Air Transport Association (IATA), which represents about 290 airlines, accounting for 82 percent of global air traffic. IATA has already made some headway in global travel and tourism standards, for example, by introducing electronic tickets and e-boarding passes which are used all over the world. Since 2020, the IATA Timatic team has been working to collect ever-changing travel regulations in over 220 countries from 1,700 government sources such as official organizations ranging from immigration departments, foreign ministries, and civil aviation authorities to ministries of health. This entire information is uploaded onto an online platform, which contains a free interactive world map to provide travelers with the latest COVID-19 entry regulations by country. It also contains complete information on the documentation required for international travel.

To keep up with the dynamic situation with respect to COVID-19, the Timatic platform is updated more than 200 times a day to provide precise travel restrictions specific to the current pandemic, based on citizenship and country of residence.

3.3 Other services approaching the new normality

As is well known, Web 2.0 can be considered as a real container of information about the world, accessible by all, sought by Internet users based on their expectations and interests and no longer (only) limited to communicators. In actual travel practices, the Internet, along with other mobile communication technologies, is becoming more and more ubiquitous (e.g., Urry, 2002; Molz, 2004; Sheller and Urry, 2006).

For example, through its blog (www.blog.google.com), Google has announced that since the pandemic broke out, more than 1 billion people have consulted Google Maps to move into the new normal. In fact, Google Maps embedded a series of COVID-related functions right from the start, helping users to obtain essential information ranging from the coronavirus situation, safety regulations, public transport (buses, trains, subways), the opening hours of public services, to crowding forecasts. Through a real-time feedback system, citizens and visitors have the opportunity to identify the best times to visit the places they would like to go. At a later stage, Google Maps added further functions within a specific section of the app called "COVID-19 info." This contains the ratio of new positive cases per 100,000 people found in one specific area, as well as an indication showing the trend based on information available nationwide.

In addition, a scale of different colors makes it possible to identify the intensity of pandemic cases in a specific geographical area, from gray, which marks less than 1 case, to dark red, which identifies more than 40 registered and suspected cases. As the area of interest narrows, the information is gradually shown at a national, regional and provincial level, to that relating to individual cities, with coverage of the over 220 countries under the census of the Google Maps service.

The information used by the service comes from institutions and accredited sources that already disseminate data on the global situation of the pandemic, such as the World Health Organization, the ministries of health of various countries, hospitals, as well as other health institutions, national and local bodies, thereby providing potential tourists with increasingly accurate and updated data in real time. Other minor innovations concern route planning to medical facilities and centers for COVID-19 testing, and now they contain more simplified and useful information to avoid an overloaded of national health services, and new alerts, in case people have to cross the border to reach a destination.

Google developed an additional application called Sodar, which can be used online, showing exactly how far people have to stay away from others in order to maintain physical distancing. All the experts indicate that people must stay at least 2 meters away from each other to limit the chances of contagion and virus spread. Sodar is applicable to the Chrome browser on mobile devices such as smartphones and tablets. After calibrating the camera, Sodar uses augmented reality, which is a layer of information superimposed on a display that shows images of the real world. More specifically, using the WebXr standard that Google has implemented on the Chrome browser for Android, it is able to represent the diameter of a circle on the video space that indicates quite precisely where the two meters end. This tool can be useful when people are in a crowded place or they have to queue up.

3.4 The algorithm of decision support for travelers

To support travelers who intend to face or placate their abroadphobia, in addition to interactive maps, algorithms have been created by different private companies. These combine quantitative and qualitative sources to indicate which territories have (or have not) managed to contain the spread of the virus. Many of them rely on artificial intelligence (AI) and big data technologies.

Among the first COVID-19 control systems designed for travelers, the "COVID controls" website (COVIDcontrols.co), was created by a team from the MIT in Boston. This allows people to gain access to constantly updated information for every place in the world, thereby enabling them to know exactly what is happening, but also which are the safest destinations to visit. On the homepage there is a map showing detailed news divided by categories for each destination. The "Lockdown Measures" section contains the updated quarantine measures imposed and lists the planned openings, for example, of restaurants, tourist attractions and shops. Similarly, the "Tourist Restrictions" section contains a list of all the measures that each country has provided for travelers: opening or closing borders, security protocols including but not limited to serological tests, swabs or isolation to be undertaken on arrival. There are also specific sections that contain information relating to the pandemic, such as "COVID controls" indicating how each country managed to contain the virus, the growth of cases based on the trend of the contagion curve of the previous 15 days, death rate, death toll and latest government announcements regarding containment measures. Through a sophisticated AI system, the data on the site are collected in an automatic manner from over 500 official sources among the ministries of health and tourism institutions of over 220 countries. COVID controls is integrated into the section "Escape" and allows people to set the search for travel only in countries where the lockdown has ended and the situation of pandemic had been brought under control.

Similarly, it is important to add that the first European Online Travel Agency, eDreams ODIGEO, has launched an AI platform for the analysis of travel trends, creating an indicator for destinations in accordance with demand trends. The service provides quality insights into global industry trends and traveler habits and seeks to support national and local tourism bodies and travel decision makers in the development of effective strategies. The group uses an advanced set of algorithms to process anonymized data from more than 21 billion annual online travel searches and to provide real-time information.

eDreams ODIGEO was also followed by the Lastminute.com group, which in 2020 launched the new "Travel Safe Algorithm" that scans destinations and analyzes them from different angles. In particular, the algorithm draws a full and detailed profile of each destination, demonstrating data ranging from the spread of the virus, the quality of healthcare, the number of UNESCO World Heritage Sites, to accommodation facilities. Every week the Travel Safe Algorithm draws up a ranking, which, excluding the countries that have closed borders, indicates the safest destinations for travelers. It considers only countries but not individual regions or cities because specific data is less reliable than national data. The algorithm does not just monitor the current situation but studies the likelihood of pandemic developments.

A similar idea was also developed by the international search engine Skyscanner. In November 2020 it created a COVID-19 interactive travel map that shows the restrictions in force to enter every country. The user must indicate the place of departure and the destination to know the most up-to-date information on quarantine restrictions, the rate of coronavirus cases and its week-over-week change. It is in beta version and can be set to different countries of origin to display different restrictions. The tool uses International Air Transport Association (IATA) data for travel restrictions in combination with Johns Hopkins University coronavirus data for case rates.

Finally, Sabre and Google Cloud joined forces in January 2021 to predict, develop and provide features that can help the evolution of the travel ecosystem. These two giants are partnering for an AI platform that represents an absolute novelty in the travel sector. The project, under the name of Sabre Travel AI, is a combination of Google's cloud infrastructure, AI and machine learning capabilities with Sabre's deep knowledge of the travel industry. The project has been designed to create third generation solutions that are smarter, faster and more cost-effective. Airlines, travel agencies, large corporations, food and drink services and other travel industry partners—offering the most relevant travel option, at the most appropriate time and on all relevant channels—are placed together to provide travelers tailored personalization, lead to higher conversion rates and increase customer loyalty. In addition, to promoting their offers more effectively, Sabre Travel AI, will enable tourism companies to increase distribution strategies across all channels in a uniform approach, and also launch guided solutions for the airports, and mobile applications on the market. For the future, Saber and Google Cloud will expect tourism companies to integrate their solutions with the new Travel AI, and differentiate within the marketplace. The system will in fact provide them access to advanced technological tools by means of cloud computing which allows to prepare and store data, expand their contents by customer or third-party databases, quickly test, develop and make machine models available, learn and understand the performance of these models as well as optimize the solutions to be launched on the market quickly and easily within a modular environment.

Local solutions have also been formulated. For example, as regards Italy, Outcrowd, a startup still in its founding stage, has enhanced and perfected a completely automatic, digital solution which is AI-based and data-driven, designed specifically for destination management. The service locates the ideal destination for traveling among historic Italian villages, small towns, mountains, the seaside and lakes far from the mainstream scenic spots. At present, the data for over 600 locations have been entered on Outcrowd. The system suggests destinations that are usually rarely visited at all times of the year, verifying the attractive potential of each place and giving the traveler guidance in the final choice of her or his destination, booking included. Each location is graphically rated on a scale of 1–5 which indicates the level of crowding. It is sufficient to navigate on the map in the "Explore" section and understand which cities or places to avoid in a given period and consequently focus on nearby alternative solutions. More specifically, the algorithm calculates the tourist pressure of each single destination on a monthly basis, processing both static data (using the dataset of ISTAT, the Italian National Institute of Statistics as a historical source) and dynamic data (derived from the search habits of web users) ranging from reviews on TripAdvisor to the reports of the Italian Touring Club and Lonely Planet about the accommodation capacity, the presence of cultural and other attractions, the number of nights spent by tourists, the local population, and so on. The creators of the platform had already begun to study and analyze public ISTAT data on the number of tourists (both foreign and Italian) who visited municipalities throughout Italy every year since 2015. The goal was to offer a technological service to the market to counter overtourism. The outbreak of the pandemic, however, made it even more intelligent to address the issue of alternative destinations and the possibility of directing tourist flows, convincing the Outcrowd creators to make their service available online.

Conclusions

The technological tools that have been presented in the previous pages can be considered as keys to orient travelers in the evaluation and choice of destinations in an innovative way. Compared with in the past, people need to keep tabs on a number of important variables before traveling.

The contemporary tourism competition is deeply "covidized" on a global level (Pai, 2020), namely, conditioned by the coronavirus and its penetration, evolution and change. The use of this neologism and other words such as "quarantine," "isolation," "outbreak," "COVID-free zone" present a sociological indicator of the change seeping into daily life in general and into the tourism field in particular.

Since the outbreak of the pandemic, an important milestone is the availability of COVID-19 vaccines, which can make a difference in controlling and containing the pandemic. However, policies on vaccinations vary across the world, and vaccination campaigns has been marked by delays, shortages and bureaucratic errors in many countries. Since epidemiologists are also worried about the fact that the spread of new virus variants may expand the infection, it follows that travelers must be even more cautious and informed in planning their travel. They must not only plan international travel as mentioned earlier, but bear in mind that this situation will probably not change in the near term. Moreover, whether traveling domestically or abroad, safety should always be a top priority. Even within the same country or region, there may be areas that are more exposed to risk than others.

The tools discussed in the previous pages make it clear that there is an analytical division of the world into constituent areas according to mobility and the predisposition to social interaction. In fact, these tools not only guide travel choices, but also condition the behaviors that travelers have to assume, or at least some of them. From this critical angle, it is clear why places and physical distances cannot be considered the determining factors of proximity or distance. Thus, the current fragmented and complex scenario is giving shape to a sort of ecological reconfiguration, making tourists aware that distances are undergoing a redefinition in contemporary society. Tourist distance today is completely different from the traditional spatial distance that separates destinations. Indeed, the restrictions imposed on mobility have helped to render a new meaning to remoteness and proximity. From a tourist point of view, today the territories that can be considered closest—even if they are many kilometers apart in physical space—are the most easily accessible because they are characterized by low virus penetration, open borders or no large-scale precautionary measures, such as a long quarantine. In other words, social actors today interpret space, distance, accessibility to the world on the basis of how territories are represented in relation to safety and governmental rules.

A clear example of this situation is represented by the travel conditions for Italians, laid down by Prime Minister Mario Draghi under the emergency decree that came into force on March 6, 2021 till April 6, 2021. During that period Italians were forced to remain in their towns (in "red zones") or their region (in "orange zones"), but at same time the Italian Interior Ministry confirmed that people were free to depart to other countries for tourism within the EU or Schengen Zone, also traveling within Italy to reach the airport or a ferry terminal. This apparently anomalous situation, together with all the other similar ones, offers the possibility of a re-reading of the sociologist Simmel's idea in a postmodern key: he defined space as something perceptive (Simmel, 1904). In fact, he argued that the perception and evaluation of spaces, which originates in the human psyche, constitutes a sociological function capable of animating and conditioning relational dynamics. The apparent neutrality of spaces is actually conditioned by the process of association (*Vergesellschaftung*). Therefore, the neutrality of physical spaces would actually be spurious, as the sociologist observed, even empty space reveals itself as the bearer and expression of a sociological action. Simmel also spoke of an "in-between" (*das Zwischen*), an intermediate entity that overlaps the physical space among people. This is a function that allows parties to meet each other. In his view, social relationships based on proximity exist, and so do others based on distance, but this distinction is never clear or definitive. In summary, it is not spatial proximity or distance that creates the particular phenomena of vicinity or extraneousness (e.g., Gieryn, 2000; Chiesi, 2010; Löw, 2016). Simmel underlined that modernity has marked the transition from an essentially topological way of relating to space, based on qualitative perception and identification, to a Euclidean-type spatiality, centered on cognition and an abstract and undifferentiated identification that transcends the space-time frame of experience, to recreate levels of experience that are less dependent on the immediacy of the latter (e.g., Mandich, 1998; Frisby, 2002).

Therefore, by translating the discourse to contemporary mobility, a territory can be considered as being close or reachable only when it makes social interaction possible. At present, the tourist distance among destinations also has a structural connotation, which concerns the division of territories into unequal areas, characterized by the level of danger. In other words, the geography of contemporary tourism allows us to consider destinations that are spatially distant as being closer or, on the contrary, consider adjacent territories or in any case not so physically distant places as being far away, if they are cut off from the global tourism market.

These statements highlight the complex network of interrelations between spatial distance and a more accurately sociological conception of distance.

Furthermore, according to Simmel, we can argue that borders are fluid, precisely because they are arbitrary. The characteristics of these configurations are inextricably linked to social connections since the sociological meaning of space is intertwined with that of relationships. Consequently, even distances today more than ever have to be considered changing. The pandemic has transformed the quality of spatial boundaries, which are becoming more and more evanescent and changeable, into an even narrower definition. Their mutation is closely related to the evolution of the pandemic in a relational and relative sense.

As we will see in more detail in the last chapter of this volume, the space-time compression made possible by the increase in new communication technologies also allows us to break through limitations, freeing mobility from geographical restriction. Speaking of proximity and distancing in reference to physical distances appears obsolete in contemporary society (Giddens, 1990).

However, the need to physically move and visit places in real person remains an aspect that does not diminish in terms of its importance in our societies (e.g., Boden and Molotch, 1994; Collins, 2004; Ling, 2008). In the covidized social scenario this topic is re-emerging powerfully. It has manifested itself even more vigorously following the many lockdowns that have been imposed. These measures have accentuated the desire to visit other places which are nearby from a tourist's point of view as well as territorially, and other locations which are tourists' own cities (e.g., American Express Travel, 2020).

The great variety of rules and limitations from country to country, to which those decided by individual transport companies are often added, makes it difficult both to manage travelers at an international level and to understand the behavior to be taken by tourists. As will be explained more deeply in the conclusions of this book, these circumstances have stimulated a fruitful discussion on the institutionalization of some documents issued by authorized health authorities or by private health companies. These "immunity passports" certify that whoever has received the vaccine or becomes immune after the viral infection is able to travel abroad more easily.

Beyond the questionable aspects that can be connected to a digital passport (relating to privacy or limitations on individual freedoms), these immunity passports could however constitute a single, formal and standardized source for providing the necessary information to travelers from all over the world. In other words, their possible additional function could be to configure themselves as a unique tool that, in addition to being a means of collecting information on the health status of their owners, would also help them to orient themselves around different tourist destinations and on the measures to be taken in order to travel to the many territories of the world. In this sense, the immunity passport should therefore also contain information on the entry rules adopted by individual countries, thereby freeing tourists from the laborious search for additional information when organizing trips.

Bibliography

Abd-Alrazaq, A, Alhuwail, D, Househ, M, Hamdi, M and Shah, Z (2020) 'Top concerns of tweeters during the COVID-19 pandemic: Infoveillance study', *Journal of Medical Internet Research*, 22: e19016.

Abdullah, M, Dias, C, Muley, D and Shahin, M D (2020) 'Exploring the impacts of COVID-19 on travel behavior and mode preferences', *Transportation Research Interdisciplinary Perspectives*, 8, 100255.

American Express Travel (2020) *Amex Trendex*, New York: American Express Travel.

Arpaci, I, Alshehabi, S, Al-Emran, M, Khasawneh, M, Mahariq, I, Abdeljawad, T *et al.* (2020) 'Analysis of twitter data using evolutionary clustering during the COVID-19 pandemic', *Computers, Materials and Continua*, 65: 193–204.

Banerjee, S (2020) 'Navigate safely with new COVID data in Google Maps', *Google Blog*, 09/23/2020.

Bargain, O and Aminjonov, U (2020) 'Trust and compliance to public health policies in times of COVID-19', *Journal of Public Economics*, 192: 104316.

Bell, D (1973) *The Coming of Post-Industrial Society: A Venture in Social Forecasting*, New York: Basic Book.

Berman, J and Bruckman, A (2001) 'The touring game: Exploring identity in an online environment', *Convergence*, 7, 3: 83–102.

Boden, D and Molotch, H J (1994) 'The compulsion of proximity', in R Friedland and D Boden (eds.) *Nowhere: Space, Time and Modernity*, Berkeley: University of California Press.

Boon-Itt, S and Skunkan, Y (2020) 'Public perception of the COVID-19 pandemic on twitter: Sentiment analysis and topic modelling study', *JMIR Public Health Surveillance*, 6: e21978.

Brainard, J (2020) 'Scientists are drowning in COVID-19 papers. Can new tools keep them afloat?', *Science*, 05/13/2020.

Bruns, A, Harrington, S and Hurcombe, E (2020) 'Corona? 5G? Or both?: The dynamics of COVID-19/5G conspiracy theories on Facebook', *Media International Australia*, 177: 12–29.

Campbell, C (2005) 'The craft consumer: Culture, craft and consumption in a post-modern society', *Journal of Consumer Culture*, 5, 1: 23–42.

Chande, A, Lee, S, Harris, M, Nguyen, Q, Beckett, S J, Hilley, T *et al.* (2020) 'Real-time, interactive website for US-county-level COVID-19 event risk assessment', *Nature Human Behaviour*, 4: 1313–1319.

Chiesi, L (2010) *Il doppio spazio dell'architettura. Ricerca sociologica e progettazione*, Naples: Liguori.

Choen, E (2004) *Contemporary Tourism. Diversity and Change*, Kidlington: Pergamon.

Citroner, G (2020) 'Travel Plans? This Interactive Map Will Show Your Risk for COVID-19', *Healtline*, 11/16/2020.

Collins, R (2004) *Interaction Ritual Chains*, Princeton: Princeton University Press.

Corbisiero, F and Monaco, S (2021) 'Post-pandemic tourism resilience: Changes in Italians' travel behavior and the possible responses of tourist cities', *Worldwide Hospitality and Tourism Themes*, 13: 3: 401–417.

Council of the European Union (2020) 'COUNCIL RECOMMENDATION (EU) 2020/1475 of 13 October 2020 on a coordinated approach to the restriction of free movement in response to the COVID-19 pandemic', *Official Journal of the European Union*, 10/13/2020.

Council of the European Union (2021) 'Update to the Council Recommendation on a coordinated approach to the restriction of free movement in response to the COVID-19 pandemic', *Official Journal of the European Union*, 01/25/2021.

Dandapat, S, Bhattacharyya, K, Sai, K, Saysardar, K and Maitra, B (2020) 'Impact of COVID-19 outbreak on travel behaviour: Evidences from early stages of the pandemic in India', *SSRN Electronic Journal*, 9.

Dann, G N S (2002) *The Tourist as a Metaphor of the Social World*, Oxon: CABI.

de Campos-Rudinsky, T C and Undurraga, E (2021) 'Public health decisions in the COVID-19 pandemic require more than follow the science', *Journal of Medical Ethics*, 47, 5: 296–299.

De Masi, D (1985) *L'avvento del post-industriale*, Milan: Franco Angeli.

Featherstone, M (1991) *Consumer Culture and Postmodernism*, London: Sage.

Fetzer, T R, Witte, M and Hensel, L (2020) 'Global behaviors and perceptions at the onset of the COVID-19 pandemic', National Bureau of Economic Research, 27082: 1–47.

Frisby, D (2002) *Georg Simmel*, London: Routledge.

Giddens, A (1990) *The Consequences of Modernity*, Cambridge: Polity Press.

Gieryn, T F (2000) 'A space for place in sociology', *Annual Review of Sociology*, 26: 463–496.

Gmelch, S B (ed.) (2004) *Contemporary Tourism. Diversity and Change*, Oxon: Pergamon.

Holt, D (1997) 'Poststructuralist lifestyle analysis: Conceptualizing the social patterning of consumption in modernity', *Journal of Consumer Research*, 23: 326–350.

Jimenez-Sotomayor, M R, Gomez-Moreno, C and Soto-Perez-de-Celis, E (2020) 'Coronavirus, ageism, and twitter: An evaluation of tweets about older adults and COVID-19', *Journal of the American Geriatrics Society*, 68: 1661–1665.

Kaewkitipong, L (2010) 'Disintermediation in the tourism industry: Theory vs. practice', in M L Nelson, M J Shaw and T J Strader (eds.) *Sustainable e-Business Management. AMCIS 2010. Lecture Notes in Business Information Processing*, Heidelberg: Springer.

Kooistra, E B and van Rooij, B (2020) 'Pandemic compliance: A systematic review of influences on social distancing behaviour during the first wave of the COVID-19 outbreak', *SSRN Electronic Journal*, 3738047.

Kreps, S E and Kriner, D L (2020) 'Model uncertainty, political contestation, and public trust in science: Evidence from the COVID-19 pandemic', *Science Advance*, 6: 43.

Kumar, K (1995) *From Post-Industrial to Post-Modern Society: New Theories of the Contemporary World*, Malden: Blackwell Publishing.

Lai, D, Wang, D, Calvano, J, Raja, A S and He, S (2020) 'Addressing immediate public coronavirus (COVID-19) concerns through social media: Utilizing Reddit's AMA as a framework for public engagement with science', *PLoS One*, 15: e0240326.

Lanfant, M F, Allcock, J and Bruner, E (1995) *International Tourism Identity and Change*, London: Sage.

Larsen, J, Urry, J and Axhausen, K W (2016) 'Networks and tourism: Mobile social life', *Annals of Tourism Research*, 34, 1: 244–262.

Lew, A A, Cheer, J M, Haywood, M, Brouder, P and Salazar, N B (2020) 'Visions of travel and tourism after the global COVID-19 transformation of 2020', *Tourism Geographies*, 22, 3: 455–466.

Ling, R (2008) *New Tech, New Ties. How Mobile Communication Is Reshaping Social Cohesion*, Cambridge: MIT Press.

Löw, M (2016) *The Sociology of Space: Materiality, Social Structures, and Action*, New York: Palgrave Macmillan.

Mandich, G (1998) 'Analogie e Metafore della Complessità: Spazio e Reti Sociali', *Quaderni di Sociologia*, 17: 147–165.

McIntosh, A, Hinch, T and Ingram, T (2002) 'Cultural identity and tourism', *International Journal of Arts Management*, 4, 2: 39–49.

Molz, G J (2004) 'Playing online and between the lines: Round-the-world websites as virtual places to play', in M Sheller and J Urry (eds.) *Tourism Mobilities: Places to Play, Places in Play*, London: Routledge.

Monaco, S (2019) *Sociologia del Turismo Accessibile. Il diritto alla Mobilità e alla Libertà di Viaggio*, Varrazze: PM Editore.

Nuryanti, W (1996) 'Heritage and postmodern tourism', *Annals of Tourism Research*, 23: 249–260.

Pai, M (2020) 'Covidization of research: What are the risks?, *Nature Medicine*, 26, 8: 1159.

Palmer, C (1999) 'Tourism and the symbols of identity', *Tourism Management*, 20, 3: 313–321.

Parmet, W E and Paul, J (2020) 'COVID-19: The first post-truth pandemic', *American Journal of Public Health*, 110, 7: 945–946.

Rosenbloom, B (2007) 'The wholesaler's role in the marketing channel: Disintermediation vs. reintermediation', *The International Review of Retail, Distribution and Consumer Research*, 17, 4: 327–339.

Rovetta, A and Bhagavathula, A S (2020) Global infodemiology of COVID-19: Analysis of Google web searches and Instagram hashtags', *Journal of Medical Internet Research*, 22: e20673.

Ryan, C and Glendon, I (1998) 'Application of leisure motivation scale to tourism', *Annals of Tourism Research*, 25, 1: 169–184.

Rzymski, P, Nowicki, M, Mullin, G E, Abraham, A, Rodríguez-Román, E, Petzold, M B *et al.* (2020) 'Quantity does not equal quality: Scientific principles cannot be sacrificed', *International Immunopharmacology*, 86: 106711.

Salazar, N B (2006) 'The anthropology of tourism in developing countries: A critical analysis of tourism cultures, powers and identities', *Tabula Rasa*, 5: 99–128.

Seytre, B (2020) 'Erroneous communication messages on COVID-19 in Africa', *American Journal of Tropical Medicine and Hygiene*, 103, 2: 587–589.

Sheller, M and Urry, J (eds.) (2006) *Mobile Technologies of the City*, London: Routledge.

Simmel, G (1904) *Kant. Sechzehn Vorlesungen gehalten an der Berliner Universität*, Berlin: Duncker & Humblot.

Stamboulis, Y and Skayannis, P (2003) 'Innovation strategies and technology for experience-based tourism', *Tourism Management*, 24, 1: 35–43.

Trauer, B and Ryan, C (2005) 'Destination image, romance and place experience: An application of intimacy theory in tourism', *Tourism Management*, 26, 4: 481–491.

Urry, J (1990) *The Tourist Gaze, Leisure and Travel in Contemporary Societies*, London: Sage.

Urry, J (1995) *Consuming Places*, London: Routledge.

Urry, J (2002) 'Mobility and proximity', *Sociology*, 36, 2: 255–274.

Urry, J and Larsen, J (2011) *The Tourist Gaze 3.0*, London: Sage.

Yamin, A E and Habibi, R (2020) 'Human rights and coronavirus: What's at stake for truth, trust and democracy?', *Health and Human Rights*, 1.

Yi, J S, Kang, Y A, Stasko, J T and Jacko, J A (2007) 'Toward a deeper understanding of the role of interaction in information visualization', *IEEE Transactions on Visualization and Computer Graphics*, 13: 1224–1231.

4 A system in crisis

Means of transport in search of solutions and new functions to withstand the pandemic

Introduction

For the protection of people's health, travel bans and other coordinated restrictions have been in operation since the outbreak of the virus. Governments around the world have been called upon to implement policies and measures to combat COVID-19. They have responded not only by introducing restrictions to commercial transport but also, at the same time, implementing temporary travel restrictions to limit the mobility of people. These measures allowed essential travel only, with very few exceptions. The tourist flows for business travel or recreational travel have been shrunk dramatically.

This situation has led to a serious economic crisis that has hit various commercial sectors including tourism.

According to the data reported by the Organization for Economic Cooperation and Development (OECD, 2020), in the first half of 2020, accommodation facilities were those most affected by the pandemic. In Europe, 76 percent of hotels have closed, shared accommodations have suffered a reduction in bookings and even Airbnb has laid off 25 percent of its workforce for the sake of survival (e.g., Boros, Dudás and Kovalcsik, 2020). In 2020, seasonal holiday destinations, such as ski resorts, ended their season earlier than expected, and so did beach resorts.

In terms of the international catering industry, restaurants initially tried to adapt to the new rules of physical distancing in an attempt to avoid the spread of the virus, but they quickly found themselves in a real economic crisis. During the lockdown businesses open to the public were forced to close completely or partially. In some cases, depending on the intensity of the virus in the individual territories, restaurants were permitted takeaway or home deliveries, resulting an evident drop in their revenues (e.g., Dube, Nhamo and Chikodzi, 2020; Siddhartha, 2020; Bucak and Yiğit, 2021; Kim, Kim and Wang, 2021; Roy *et al.*, 2021; Song, Yeon and Lee, 2021).

For this reason, a series of supporting policies and initiatives to provide liquidity have been carried out at the national and international level.

DOI: 10.4324/9781003195177-5

More specifically, task forces were created for the entire tourism system with the aim of establishing a gradual and generalized coordination that would work for the recovery of the whole tourism sector. One of these is the "World Travel and Tourism Council" which was created to relieve the pressure on the global tourism businesses by linking the private sectors with international organizations. With the same purpose, the Global Tourism Crisis Committee was established, led by the United Nations World Tourism (UNWTO). Starting from April 1, 2020, this task force published a series of recommendations aimed at mitigating the impact of COVID-19 on employment and liquidity, helping the different sectors and the most vulnerable players with recovery. Subsequently, in May, this body approved the "UNWTO Global Guidelines" to relaunch tourism nationally and internationally, indicating to the various governments the supporting initiatives to be carried out.

Governments around the world have taken the lead in implementing emergency measures and laying down possible strategies for the recovery of the tourism sector.

In addition to the economic problem, which certainly represents a very important aspect, another problem exists. People have become afraid of traveling and no longer (or not only) associate travel with the idea of a moment of vacation, fun or personal growth, instead, they see it as a chance of infection (e.g., ART, 2020; Budd and Ison, 2020; Ho *et al.*, 2020; Suau-Sanchez, Voltes-Dorta and Cugueró-Escofet, 2020; Zhang, Hayashi and Frank, 2021). The reassurances from medical experts have not been enough to dispel their misgivings. Even the new security measures implemented by some transport companies have not been able to fully reassure travelers. Some news events have contributed to fuel this phobia. For example, on September 29, 2020, flight EK448 from Dubai to Auckland experienced a small-scale outbreak of COVID-19. According to a relevant study (Swadi *et al.*, 2021), the virus spread on the plane from a virus-carrier on board whose infection was not detected or who perhaps contracted the virus during one of the stopovers.

We have already explained how the fear of traveling has damaged domestic tourism on the one hand and, on the other hand, has led to an aversion toward traveling abroad, namely, abroadphobia. This is not derived from epidemiological data, but is developed by an inadequate understanding of foreign health systems and the fear of contracting the virus in a distant foreign land.

A survey published in April 2020 by the International Air Transport Association (IATA, 2020b) found that 40 percent of respondents felt insecure about long-distance travel, considering it to pose a threat to health.

Domestic and international tourism is reliant on the transport system, which began to experience a profound economic crisis since travel bans were activated and new regulations affected the sector. In particular, some transportation junctions have been closed, infrared thermometers have been installed in airports, railway and bus stations and daily disinfection of means of transport has been provided in case of necessity.

At the European level, the European Commission announced in the first months of 2020 the closure of the external borders of the European Union, with the only exception of the so-called "fast lines," namely, preferential routes for maintaining the continuity and regularity of the freight transport of railway, sea and air. At the same time, many European member countries have issued provisions to ensure internal mobility, without disobeying the highest safety standards. It can be deduced that transport was one of the tourism sectors most affected by the coronavirus emergency, given the reduction in routes, trips and reservations. In this catastrophic scenario, some transport companies have tried to reinvent themselves, proposing alternative services. As we will see in this chapter, the coronavirus has certainly brought great damage to the sector, but at the same time it also produced novel results. These desirable results demonstrate that creating innovation in the transport sector would help combat the pandemic crisis.

4.1 Air transport

On December 17, 1903, the Wright brothers successfully managed to fly their rudimentary machine at the height of a few hundred meters. This historical day was later marked as the date of birth of the airplane (Jakab, 1990). Actually, as early as 1900, Zeppelin had built an airship, which was equipped with two internal combustion engines, making it possible to enjoy quite comfortable, albeit slow, expensive and not very safe journeys. In 1917, the United States Postal Services established the first airport service in the world. In August 1919, the first freight transport was carried out with a military bomber converted to transport heavy loads. The First World War gave a great impetus to the research and production of numerous aircraft models that were then used for war purposes (e.g., Hallion, 2003).

A major transformation of military aircraft into civil aircraft took place in the 1950s. In the 1960s, thanks to the birth and development of jet engines, the increase in the autonomy of aircraft allowed for a reduction in fuel consumption. In the following decades this led to the creation of "parcel express" services and the development of "all cargo" operators.

In America in the 1970s low-cost airlines made their appearance. These are air transport companies which offered only the indispensable and basic service of the journey, and other paid services were offered only at the request of passengers. The commercial flight market was monopolized by a few large groups which formed a real tariff cartel and therefore needed a liberalization of routes and fares.

Subsequently, in Europe too, between 1983 and 1992, through a series of regulatory packages, the deregulation of air transport was decreed, which sanctioned free competition in the skies. The Irish airline Ryanair was the first to fly to Europe after this reform, and the company still occupies the largest share of the market today. Low-cost services differ from charter flights in that the former is scheduled and not occasional, rationally organized to save money specifically so they can offer a low-cost ticket. The low-cost airlines, in most cases, do not land in the main airports, but often use secondary airports as hubs. These are relatively close to the main airports, and the costs for the air carrier are lower and these secondary hubs may have a greater number of slots available. This has led to a series of important transformations in the air transport sector.

In Europe they acquired a real success in the early 2000s (e.g., Dobruszkes, 2006; Malighetti, Paleari and Redondi, 2009; Pels, 2008; Pels, Njegovan and Behrens, 2009). Among others, the most significant changes were represented by the entry of numerous new operators on the market (especially first-generation low-cost carriers and regional carriers), and the consequent increase in competitiveness. Furthermore, since the early years of this century, tourists have been able to enjoy the choice of a variety of flights and more routes previously not served by the major companies, as well as new regional transport. Likewise, the number of international passengers has increased even more and fares have undergone a significant reduction (Arrigo and Giuricin, 2006; Francis *et al.*, 2006; Dobruszkes, 2013). The attitude of low-cost airlines has resulted in a series of unprecedented marketing actions, including offering services at very low prices and selling tickets directly to travelers, as in the e-business pattern. These low-cost airline companies have succeeded in breaking the national airlines' market monopoly, offering the tourist market access to many smaller cities previously almost excluded from the global tourist market due to inadequate transportation.

In Europe, the growth of passengers who enjoy low-cost airlines has been exponential. In 1994 about 3 million passengers chose the services of low-cost airlines and in 1999 the number rose to 17.5 million. The first decade of the third millennium witnessed the boom of low-cost airlines: for example, Ryanair alone attracted 65 million passengers in 2009 and over 72 million in the following year, registering a year-on-year growth of over 10 percent (e.g., Almeida and Costa, 2012; Diaconu, 2012).

Travelers' urges to visit different places have equipped planes with the function of socialization. Planes have become the means of transport par excellence, capable of connecting the most distant cultures by facilitating the journey. Sociologically, confrontation with different cultures in different territories serves as one of the engines of social change. It is made possible by long-distance mobility, in which airplanes play a central role because they allow geographically distant places to connect in a relatively limited number of hours, activating a concrete cultural exchange on a global level.

The increase in international flights has made it necessary to structure the "hub and spoke" of the air service. Airlines have increasingly sought to attract as many passengers on board as possible for the maximum economic benefit. For them, economical guarantee comes from stopovers, during which the comings and goings of passengers are registered and the possibility of bringing a larger number of people on board increases. Thus, all the major cities in the world have also become hubs, or to be more specific, passenger collection centers, where tourists arrive from many peripheral airports.

As we anticipated in the first chapter, the historical road of aviation has not been a smooth one. In particular, 9/11 represented a moment of great fear lingering in the minds of tourists. This event forced airlines around the world to attach absolute importance to air safety. They had to improve preflight inspections and regular maintenance of the aircraft, carrying out disassembly and reassembly of the entire aircraft at least every 6 months. In addition, airport checks on passengers have been intensely reinforced for the sake of safeguarding airport security and protecting the boarding travelers (e.g., Lyon, 2006; Blalock, Kadiyali and Simon, 2007; Peterson and Treat, 2008; Fox, 2014).

Since 2020, aviation has begun to experience a new nightmare due to the pandemic. As reported by CIRIUM (2020), the decline began with the first peak of canceled bookings in Asia in February, coinciding with the outbreak of the coronavirus in China and the closure of the borders of many Far Eastern countries. As early as February 2020, Korean Airlines, for example, declared that over 90 percent of its air fleet remained out of service at South Korean airports. As the pandemic spread to Europe and North America in the following months, the sector began to feel the blows of the global crisis more violently. In particular, in March 2020, the then-President Trump declared a national state of emergency and decided to block flights from the Schengen area, as well as from China and Iran.

In Europe, Easyjet, Lufthansa, Aegean and British Airways left most of their air fleet idle until the end of the health emergency. In addition, Ryanair canceled 80 percent of its flights as from midnight on March 18, 2020. During 2020 this airline reduced its capacity by 60 percent compared to the previous year and decided to close several bases throughout 2021, such as Cork and Shannon in Ireland, and Toulouse in France (Asgari, 2020). In April and May, daily flights were on average less than 30,000 per day, compared to over 100,000 of the same period in 2019. The day with the fewest flights ever was on April 25, when only 13,600 aircraft took off, marking 86 percent fewer than the busiest day, January 3, with 95,000 flights. Air travel experienced a slight recovery in June, largely because of the contribution of domestic flights. However, the recovery failed to last long, since in August many countries imposed the closure of borders even among areas in the same country in the face of the growing number of infections.

As a result of the coronavirus pandemic, in the United States, United Airlines and American Airlines laid off 32,000 employees in October 2020, after unanimous political agreement on the new aid package for the sector was not reached in Washington, D.C., and in September the financial aid that guaranteed the wages of workers was not renewed. According to the Air Transport Action Group (ATAG, 2021), before the pandemic air transport boasted 11 million jobs, and over 87 million in service staff, also considered the professions indirectly associated with the sector. The contribution of air transport to the world economy reached an estimated 3.5 trillion dollars, equal to just over 4 percent of the global gross domestic product. According to data by IATA (2020a), the number of aircraft passengers went from 4.5 billion in 2019 to 1.8 billion in 2020. From January 1, 2020 to December 20, 2020, 16.8 million flights were conducted, equal to 49 percent fewer than the 33.2 million in 2019. Passenger traffic dropped by 67 percent because of the restrictive measures and abroadphobia, with international flights suffering the greatest decline, equal to 68 percent in the time frame January 1, 2020–January 1, 2021, with numbers similar to those of 20 years previously. The airline companies were not able to cut their losses, and their contribution to the economy and employment has therefore been halved.

To revive the economy, some airlines have adopted alternative solutions aimed at limiting the damage. In particular, many of them have tried to attract passengers with discounts: for customers who joined the loyalty programs, they rewarded the most frequent travelers, or frequent flyers (IdeaWorks, 2020). Airline companies have put forward the idea of assigning "miles points" in accordance with the flight distance and redeeming them for new tickets as a strategy of promotion to increase customer stickiness.

Since the beginning of the pandemic, for example, in the same vein, Qatar Airways gave the new customers who had signed up to its frequent flyer program a bonus of up to 7,500 mileage points, and also awarded additional miles to the people who made bookings online.

A different strategy was instead adopted by the United Airlines company, which has significantly lowered the threshold of miles required for a ticket to encourage more passengers to redeem these miles under the "MileagePlus" loyalty program.

Companies in Europe also increased their efforts on loyalty programs at first.

According to British Airways, the members of its "Executive club" could save 50 percent of their mileage points (Avios) on some selected routes for a whole trip which originally needed full points.

The Spanish national airline Iberia and the low-cost airline Vueling adopted the same program: members of the "Vueling club," for example, could also accumulate Avios points on flights operated by Iberia, with which Vueling has an agreement of mutual code recognition. In other words, one company could carry passengers whose tickets have been issued by the other. Avios points did not expire unless the person of the program spent more than 3 years not redeeming or increasing them.

The Italian Alitalia also enhanced its loyalty program, postponing the expiry of the "Mille Miglia" points accumulated from December 31, 2020 to December 31, 2021 and it extended the deadline for using them to January 31, 2022. Furthermore, the booking of a flight on some routes allowed people to earn twice as the points that would normally be awarded.

Another example comes from France. Members of "Flying blue," the loyalty program of Air France and its partners, have been able to take advantage of a discount of up to 50 percent when booking a flight on some routes, but were also able to save up to a quarter of the miles for the flight to other destinations.

The largest program dedicated to European frequent flyers is "Miles & more," which involves over 300 partners and is dedicated to the customers of about forty airlines, including Lufthansa, Swiss and Eurowings. From 1 January, 2022, the program has been simplified by merging different points systems into one.

But the whole picture is far more than these.

In the midst of the COVID-19 crisis, in Australia, Japan and Taiwan several airlines offered passengers the opportunity to turn flights into panoramic tours. These flights departed from and land at the same airport, hence the campaign obtained its name "flight to nowhere," which was designed to show the world from above.

This is not an entirely new initiative, as some panoramic tours already existed in the past. However, with the advent of COVID, the offer has been in full swing. Among the first proposals, Eva Air offered flights of about 3 hours that flew over the most charming Taiwan and the Japanese Ryukyu islands, while passengers enjoyed an unprecedented culinary experience brought by a starred chef. A similar initiative was proposed by Royal Brunei Airlines, a company based in the state of Brunei, which offered a 90-minute flight to admire the enchanting landscapes that characterize the island of Borneo, together with a high-altitude dinner.

In Thailand, scheduled flights have turned into coffee shops, allowing customers to enjoy coffee or other beverages on board while flying over certain areas of the country. In Japan, Air Nippon Airways has begun offering customers 90-minute Hawaiian-themed flights to and from Tokyo. The Australian company Qantas came up with a new offer: 7 hours of a carbon-neutral "loop" flight targeted at people who wanted to experience the sensation of a long intercontinental journey again. During this long trip, travelers not only enjoyed the panoramic views of the Whitsunday Islands, the Great Barrier Reef, the Sunshine Coast and even Sydney Harbour, but they also had a meal prepared by Neil Perry, who is one of the most famous chefs in the country. People who participated these nonborder-crossing programs were exempt from quarantine or swab. Be that as it may, at the same time a number of precautionary measures were taken on board. For example, the central seats were blocked out to ensure physical distancing on board, sanitation was intensified and masks and sanitizers were in full supply.

The "flights to nowhere" have been designed to meet three different but closely interconnected demands. The first and most obvious purpose was to increase the economic benefit to the companies.

Secondly, these initiatives allowed the most passionate customers and also rational customers to resume traveling.

Finally, with the progress of these initiatives, pilots were able to conduct at least three take-offs in the space of 90 days, as required by the pilot license. To this end, in the first few months of the pandemic some pilots had no alternatives but to conduct pseudoflights (without any passengers on board) equipped with simulators. But this was nothing more than an expedient, given the fact that a large number of pilots had remained idle for months.

Toward the end of 2020, Taiwan airline Eva Air began organizing a kind of in-flight speed dating. The so-called "You and Me" events were carried out with a per capita "fare" of around $300. Each flight could carry a maximum of 40 passengers, of which the male to female ratio was 1:1. The passengers had to meet certain requirements: graduates and Taiwanese citizens, male aged between 28 and 38; female between 24 and 35; all seats were randomly allocated. During meals on the ground, participants had the opportunity to move around and chat with whomever they preferred in search of love. Participants were encouraged to get to know each other while enjoying the dishes prepared by a starred chef, and also to wear a mask when they were not drinking or dining. The first completely full flight took off at Christmas 2020, and it turned out to be a huge success.

In October 2020, Air New Zealand decided to offer a more brilliant service, not too innovative, but in a short time it caught the attention of lots of tourists, attracted to the chance to get back on the road. This idea was called "mystery breaks," a package of tourist services including flights, hotels and rental cars, characterized by informing travelers of the destination (within the national borders) only 2 days before departure. These kinds of experiences have been designed to encourage local tourism with new proposals tailored to every taste and pocket. The traveler knew the complete itinerary among the 20 possible routes only 48 hours before leaving. At the time of booking, they could exclude only one destination. The price of a "Mystery break" started from around $350 for a weekend and people could get as much as they wanted, depending on the package they chose. The possibilities covered Great, Deluxe and Luxury packages, which included overnight stays in five stars and luxury cars.

In March 2021, the Qantas company started offering a similar service: "mystery flights," that were a series of surprises to all people who decided to take part in the project. This has been another solution designed to cope with the decline in the number of passengers. According to the Association of Asia Pacific Airlines (AAPA, 2021), in the Asian and Australian airlines, in January 2021 only 1.3 million passengers took a plane, while in 2020, 33.5 million travelers flew to the region. A "mystery flight" is actually a trip without a destination. Its history can be traced back to the 1990s when the Australian company first started the campaign. In fact, in the 1990s, passengers who signed up for the "mystery flight" organized by Qantas were unaware of the destination not far away from home, but in the COVID-19 era this kind of experience has become truly complete, because it has been no longer exclusively about the flight experience; instead, it included surprises on board and at the destination. In other words, the three "mystery flight experiences" promoted by the Australian national airline provided in fact an all-round adventure, albeit a short one. The "mystery flights" were carried out by one of three Qantas Boeing 737 planes from Brisbane, Melbourne or Sydney with cheap fares starting at AU $737, about $577. Qantas took its passengers to a mysterious and less well-known destination on a 2-hour flight. The surprises already began during the flight, as pilots enabled the passengers to admire the landscape from above at low altitude. Once landed, on the other hand, travelers had the opportunity to gain unique tourist experiences during the day, and they returned home at night. The package included various activities ranging from wine tasting, gourmet lunches, to scuba diving on tropical islands. The airline companies advised the articles of clothing to wear and gave some tips about the unknown destination so that tourists could prepare their luggage better.

From the end of 2020, Quantas had also put on sale packages containing snacks, hand cream and the airline's popular pajamas, which are normally intended for first-class travelers.

Months later also other airline companies started to sell some of the products usually offered on their planes. For example, the Canadian Air North and Singapore Airlines sold the meals they usually served during their flights, through a home delivery service. In addition, Singapore Airlines offered to nostalgic travelers the opportunity to have a meal in first or economy class on board on an Airbus A380 parked at Singapore-Changi airport.

Similar services were offered also by British Airways in 2021. The company not only put up on sale its plates, glasses, cups and the other items usually used on board, but starting from March it started to deliver a meal kit containing ingredients and instructions for preparing the typical meal that was served on a first class flight. The service has been managed by DO & CO, the catering company of British Airways, in collaboration with Feast box, in charge of the creation and distribution of the kits. To encourage British people to travel, customers who subscribed to the service could enjoy a 10 percent discount on the purchase of a future air travel with British Airways.

Finally, in Australia, a plane was chosen as a means to host the Mardi Gras 2021 celebrations in March.

In fact, Australia was the first continent to celebrate a victory against the virus, registering zero cases. Unlike what has happened in other countries, during the most critical juncture of contagion all Australian States (South Australia, Western Australia, New South Wales, Queensland, Tasmania and Victoria) totally or partially closed their borders, significantly limiting both the interstate and intrastate movement of travelers. The key to Australia's success was not only closures and lockdowns, but also the strengthened tracking system used to trace those in contact with the positive cases to circumscribe the outbreaks in the shortest possible time. Consequently, the so-called "Pride flight" was a festive occasion not only to celebrate the pride of the Australian LGBT+ community, but also to celebrate this great anti-pandemic milestone. The "Pride Flight" traveled from Brisbane to Sydney, hosting over 120 passengers, including many famous advocates of social inclusion and members of the Australian drag "royalty." The airline Virgin Australia distributed Streets Golden Gaytime ice creams to all guests prior to boarding, while team members lined the boarding queue waving rainbow flags. Unlimited beverages and food were served on board. Over the course of an hour and a half, passengers took part in several in-flight games, including seat-pocket bingo. Before landing, the flight concluded with a drag performance.

4.2 Rail transport

Although the use of steam as a source of propulsion can be dated back to Hellenic times, vehicles with steam engines did not appear until the late 1700s. The first steam locomotive was produced in 1804 in the United Kingdom, followed by the completion of the first public railway in history in 1825 in the United Kingdom too (e.g., Jones, 2012; Pirie, 2014).

Rail travel was the first type of travel that allowed mass movement of people but above all made it possible to transfer raw materials and goods of all kinds in a more economical and faster manner than hitherto.

Due to this, rail transport contributed significantly to the development of trade, occupying a central position in the transport sector at least before the extensive diffusion of cars.

During the years between the early twentieth century and the First World War, however, numerous railway companies, especially those situated in Europe, were clearly suffering from the financial crisis brought about by the start of nationalization and the large-scale subsidization by government finance on railway systems and metropolitan railways (e.g., Stevenson, 1999; Roth and Jacoli, 2013).

In the period between the two wars, the railway network was strengthened also thanks to the construction of high-speed trains (Kobayashi and Okumura, 1997; Bruinsma *et al.*, 2008; Calle-Sánchez, 2013). The first electrified high-speed rail Tōkaidō Shinkansen was introduced in 1964 between Tokyo and Osaka in Japan.

Nowadays, in the United States and many other countries, the quality of public transport often depends solely on public funding (e.g., Mulley and Nelson, 2009; Buehler and Pucher, 2011). In Japan, however, Japan East Railways manages to produce substantial profits in the public transport sector. In various European countries, in recent times, the progress of high-speed transport and the reorganization of highly specialized freight transport have been showing that profit-making is possible, especially since the separation of infrastructure management from real service. More precisely it has allowed the separation of the relative balance sheets as well as the possibility of privatizing the sector of rail transport for the sake of profit (e.g., Priemus and Konings, 2001).

As regards safety, due to its characteristics, rail transport is considered by travelers as one of the safest means of transport as it follows its own track, so normally two trains will not meet head-on or collide with the only exception of at level crossings, where road and rail cross.

Furthermore the rails ensure that trains run smoothly and also can prevent derailment, whereas going off the road is a common cause of accidents in road traffic. The possibility of multiple locomotives colliding is greatly prevented by safety systems and rules. Likewise, the driving and control of rail traffic is generally entrusted to highly qualified personnel, who are constantly updated and whose health condition is monitored. The infrastructure and railway vehicles are also subjected to periodic checks, preventive maintenance and regular replacements.

An important aspect, in rail transport as in other forms of transport, is that of precautions against terrorist events (e.g., Plant, 2004; Ciganik and Balasicova, 2008; Lieberman and Bucht, 2009). Many railways are substantially investing in raising current security standards against terrorism, although this is more difficult than for other forms of transport because stations are designed for easy access and high capacity. In addition, commuter and subway trains stop very often, making them difficult for travelers to control. On the other hand, the centralization of the spaces open to passengers and their closure allows relative control in the main stations since these are most subject to this type of attack.

To prevent the spread of coronavirus, several measures relevant to the global rail transport sector have been in force for the maximum protection of the health of travelers and workers. Regardless of the territorial specificities, some common elements can be identified: greater disinfection of trains; disinfection supply; shortened cleaning intervals; greater cleaning and disinfestation of stations, as well as common areas and workplaces in the stations; installation of hand sanitizer dispensers on board; staff provided with protective kits (masks with filters, disposable gloves, disinfectant gel); introduction of innovative online compulsory booking systems; obligation of physical distancing among passengers and compliance with the main anti-contagion measures prescribed by the provisions on the prevention of the COVID-19 virus. Specific task forces shoulder the responsibility of passenger regulation by checking the temperature of the passengers and refusing entry of those who come from the countries whose borders are closed.

Understandably, the railway companies have also gone through a period of financial crisis as a result of the pandemic. For example, during 2020, European companies lost 26 billion euros compared to 2019 (CER, 2021). Passenger services made up the largest part of the losses, recording a year-on-year decrease of 42 percent, the equivalent of 24 billion euros.

Freight transport services, on the other hand, underwent a slightly lower loss of 12 percent, the equivalent of 2 billion euros.

According to the China State Railway Group (2020) statistics, the total revenue of China's railway in 2020 amounted to 1.13 trillion RMB yuan, which has been basically equal to the previous year. However, 0.65 trillion RMB yuan in the total revenue came from transportation, and this proportion decreased by about 20 percent, compared with 2019. In 2020, the annual dispatched number of passengers reached 2.16 billion, showing a decrease of around 40 percent, equivalent to 1.41 billion, compared with 2019.

The rate of vaccination and the resumption of travel have already boosted the confidence of passengers who have been longing for a rail trip.

The most striking example of this comes from China, where passenger travel and rail freight volumes both experienced growth during the 2021 Spring Festival travel rush (China State Railway Group, 2021). In fact, from January 28 to March 8, Chinese railways handled 218 million passenger, registering an increase of 3.5 percent compared with the Spring Festival travel rush of 2020. Data from China Railway indicate that the railways transported 517 million tons of freight over the same period, up 18.1 percent from the previous year.

Meanwhile internationally, many railway companies began to offer a "COVID-free solution." A COVID-free train, as the name suggests, were a train open only to passengers who were virus negative, which could be considered a kind of free zone where people could travel in complete safety. The main feature of COVID-free trains was that only passengers who had a negative result in the nucleic acid test carried out before departure could board.

In addition to the idea of the virus-free train, two other initiatives, also associated with the pandemic, were launched in some European countries (the first were Germany, France and Italy), which experimented with a re-functionalization of trains.

The first of these has been the creation of what can be defined as "hospital trains." These were special convoys which had the function of helping to transfer people with positive COVID-19 test results—patients or other sick people—to European hospitals, and they can be considered as tools to support some health services. The idea has been inspired by hospital trains set up in Europe during the First World War. These trains had medical personnel on board to help and support travelers during the journey.

Another idea in service in some territories has been the creation of some temporary vaccination hubs in city stations. Rome Termini Station is a good example of this. Starting in March 2021 it set up a hub covering an area of 750 square meters with an emergency tent on its flanks. The Termini hub has been equipped with 17 emergency stands and 24 dedicated to vaccination, and there were also two stands exclusively used by people with disabilities.

Since the authorities in some countries were pushing to discourage the use of planes for shorter journeys by sponsoring rail transport, we can hypothesize that high-speed railway could replace short-haul flights. With reference to this in Germany, the Lufthansa airline has decided to carry out an experimental project which could also be followed in other territories by other airline companies: starting in July, 2021 Berlin, Hamburg, Bremen, Münster and Munich have been connected to the Lufthansa hub in Frankfurt via high-speed rail services, in collaboration with Deutsche Bahn (the German state railway service), with the forecast aim to cover more German routes and cities.

Finally, we must say that travel on wheels could be further strengthened by the application of innovative, high-performance and sustainable technologies. At present, for example, several investors around the world are showing interest in a project by the founder of Tesla and SpaceX, Elon Musk, who has patented Hyperloop, a magnetic induction pod which travels through a large vacuum tube at up to 1200 kilometers per hour. Its first test with passengers on board was carried out in November 2020, when it traveled 500 meters at the speed of 400 kilometers per hour in the Nevada desert.

A subsequent experiment was carried out in a dedicated area in West Virginia. However, the project needs more testing and huge funding before it becomes operational on American soil. Hyperloop routes are being planned not only in America, where the track specially for a 30-minute trip from Chicago to Pittsburgh is already under construction, but also in India and the United Arab Emirates, where it will be possible to travel from Abu Dhabi to Dubai, the first tourist destination in the world with Hyperloop, in only 12 minutes.

Important developments will also take place in Europe, with Italy among the first countries of the old continent to offer short routes, such as Milan-Malpensa and Rome-Fiumicino, in 2 minutes. Also on the cards for Italy, is a project to build a large subway in the northern part of the country, a system of tubes that connects the city of Turin with Venice, with intermediate stops. It would be possible to travel from Milan to Rome in the Hyperloop capsule in less than half an hour.

Evidently, this is a complex project, but it will exert a positive impact on the tourism sector. If travel times become shorter, it is possible to imagine new travel packages that will be even more attractive. Certainly, Hyperloop will represent a paradigm shift in the passenger experience, making it extraordinary.

4.3 Sea transport

In 1783 Clermont, a pioneering steamship, set sail along the Hudson River.

However, it was not until the end of the nineteenth century that we witnessed embryonic cruise tourism, thanks to shipping companies which set up a complicated network of sea trips, allowing European travelers to explore the New World. Thus the concept of a cruise was started, which spread with the transformation of ships into real floating hotels offering tourists comfort and hospitality (Löfgren, 1999).

Over time, cruise ships have captured an increasing interest from tourists. In fact, on a global level, the number of cruise passengers per year went from 2 million in the late 1980s to 30 million in 2019, accompanied by a growth that affected not only the richest countries, but also underdeveloped or developing ones. The demand for cruises has steadily increased, maintaining a positive trend against all the crises that the world has experienced during the past decades (e.g., Sklarewitz, 1991).

However, the arrival of COVID-19 has changed this favorable situation. In fact, the cruise, compared to other forms of travel, faces a series of additional problems. First, it involves the crossing of numerous borders during its voyage. As bans and closures have been imposed among the different areas of the world, defining an itinerary has become more complicated. A further uncertainty arises from the necessary reduction in the number of passengers to comply with rules of physical distancing, which makes the business less economically beneficial and sustainable.

Finally, the maritime transport sector has also been crippled by enormous attention from the international media as some cases of COVID-19 contagion spread on board cruise ships in the months of March and April 2020. A cruise holiday in fact includes the sharing of public space, which increases the risks of crowding. The UK-registered Diamond Princess was the first cruise ship that experienced a major outbreak of the virus on board, and consequently the ship was quarantined for about a month in Yokohama from February 4, 2020, resulting in over 700 confirmed cases and a death toll of 12 (e.g., Nakazawa, Ino and Akabayashi, 2020; Rocklöv, Sjödin and A Wilder-Smith, 2020; Yamahata and Shibata, 2020).

Moreover, in May 2020, confirmed positive cases of COVID-19 were found on board more than 40 cruise ships, notably on the Artania, which remained at sea with eight passengers on board who disembarked in Bremerhaven, Germany, about a month later (e.g., Teberga de Paula and Merlotti Herédia, 2020; Zhang and Wang, 2021).

Faced with such trying circumstances, several countries have placed limitations on docking to prevent the spread of the coronavirus, even though there has not been a concerted action among these nations (e.g., Choquet and Sam-Lefebvre, 2021). In fact, globally, under uniform regulations, maritime transport has often been affected by the failure to reach various goals. The regulations imposed by the Athens Convention of December 13, 1974, subsequently amended by the London Protocol of November 19, 1976 and March 30, 1990, and, most recently, of November 1, 2002, despite having entered into force on April 28, 1987, have to date been ratified by only a fairly small number of states. For example, since March 15, 2020, Australia has prevented all overseas cruise ships from docking, and on March 27, the government ordered all foreign ships to leave the country. On May 20, 2020, the Australian Health Minister extended the shutdown on cruise ships with more than 100 passengers, until September 17, 2020 (e.g., Moloney and Moloney, 2020; Quigley *et al.*, 2020).

Similarly, Canada prohibited all ships carrying more than 500 people from docking from March 13 to July 1, 2020.

In New Zealand, at the beginning of the pandemic, only ships already present in the country's waters were authorized to complete their itinerary. Since March 2020, the ban has been applied to all ships there.

The Seychelles government has even suspended all authorizations for cruise ships until 2022.

In South Africa, during the period of the lockdown, the closure of both land and sea borders was announced, specifying that the closure of ports did not affect only the transport of essential goods.

In this scenario, the International Maritime Organization (IMO, 2020) issued a circular encouraging governments and national authorities to ensure that ships and ports remained operational, in order to maintain the full functionality of supply chains and the circulation of maritime personnel.

At the same time, in order to get through the emergency phase, the European Commission also prepared and published a series of guidelines on recommendations on health, return and travel arrangements to support cruise passengers infected by the virus, as well as to regulate the transfer of essential goods and services for European countries through ports (European Commission, 2020).

The Europe Union requested the Member States to facilitate the return of European citizens and third-country residents in Europe. Furthermore, Europe indicated it would guarantee maritime personnel engaged in the market the opportunity to reach embarkation ports and also to disembark at the end of their missions to return home. Finally, the Commission called on EU Member States to set up a network of safe ports where ships could dock.

Similarly, Japan earmarked approximately 60 million Japanese yen for research into developing a set of international standards to tackle outbreaks of infectious diseases on board cruise ships, requesting the support of several international bodies such as the International Maritime Organization and the World Health Organization.

Although in 2021 cruise ships in Europe and Asia have resumed service with strict safety rules, tourists still have limited confidence in this form of travel, so much so that some studies foresee the almost zeroing of the operators' revenues in the short-to-medium term.

Despite the difficulties, some cruise companies have also tried to offer alternative services in addition to traditional travel, in compliance with anti-COVID rules.

In particular, Royal Caribbean International and Genting Cruise Lines, two of the largest cruise lines in the world, were among the first companies to borrow the idea pioneered by airlines of offering trips to nowhere. In practice, they devised a program in which passengers had no authorization to disembark at any port but with access to the ship's facilities, from the restaurant to the pools on board, making a journey back to the port of departure.

In October 2020, the Singapore Tourism Board also launched the cruise-to-nowhere project, rejuvenating the city state's cruise industry companies. The Singaporean tourism agency has also entrusted a Norwegian company, an international leader in consulting on medical and health care activities in the maritime health sector, to establish health and safety protocols for shipping companies.

Finally, in 2021 Thailand has gradually eased its restriction measures by reopening its borders, identifying that cruise ships could function as a place where tourists could spend the mandatory 2-week quarantine period once they arrived in the country. Travelers were welcomed on board at various locations including Phuket and Krabi. They had to remain on the cruise ship for the necessary period before disembarking. Meanwhile, tourists had to wear a smart bracelet that monitored their main vital signs including temperature and blood pressure, and their location was monitored through GPS. The device could transmit information even at sea, within a radius of 10 kilometers. The government aimed to earn 1.8 billion baht (48 million euros) from tourists thanks to this initiative in a year.

Conclusions

The pandemic shock has produced structural changes in the determinants of mobility (JRC, 2020) and a sudden change in individual and collective travel habits, which poses new challenges to the regulation and governance of mobility (e.g., Paniccia, 2020) whose degree of irreversibility is still however uncertain. In the meantime, to handle the crisis, the transport system has experimented with new solutions and initiatives, trying to mediate among the needs of tourism, and the uncertain and ever-changing pandemic situation.

Time will certainly see a gradual recovery in the sector. To this end, various actions should be implemented in order to rebuild travelers' trust. In other words, transport companies must reinforce the positive aspects of transport, ensuring safety, care and hygiene on board vehicles as well as in surrounding environments, such as stations, ports or railways.

The primary challenge will be to raise the level of health safety, offering travelers new guarantees and offers, for example, by promoting air filtration systems, encouraging their compliance with new hygiene measures on board, or redesigning spaces on airplanes, ships and trains, to ensure a comfortable and safe distance among passengers. It may also be necessary to increase the frequency of departures in order to enable more people to leave, without running the risk of creating queues or gatherings before departures. If this solution were practiced, to mitigate the environmental impact, higher costs of fares could be used to promote research and development of sustainable solutions.

On these aspects, the consulting company PricewaterhouseCoopers has developed a smart mobility demand management platform (World Economic Forum, 2020) to allow travel companies to monitor their system's service conditions in real time (e.g., about position of vehicles, crowding levels, and so on) and provide accurate near-real-time evolution forecasts (in terms of estimated arrival times, load factor evolution) to allow users to plan trips with a high level of safety and reliability. This is only one of the possible solutions that can allow operators to plan their offers dynamically, and enable local government bodies to know in real time the operating status of transport networks in consultation with the transportation system.

To ignite the desire to travel, the transport sector also has to offer a new level of comfort on board, making passengers feel at home, and apply new technologies: for example, guarantee easy access to video, data and audio on the Internet at ever-increasing speed.

Furthermore, in terms of safety, today, thanks to the development of on-board sensors and the Internet of Things, it is possible to obtain huge technological advancements, such as real-time checking of means of transport. The use of technology for the management of means of transport makes it possible to organize preventive maintenance, which costs even less than corrective maintenance. Similarly, to reduce costs and recover losses due to the pandemic, robot maintenance can be another solution, which is already financed by some European programs.

Likewise, technologies could be further improved on board and in hubs to monitor the health and behavior of passengers, such as measuring their body temperature without contact or checking whether they are wearing a mask correctly. As I have anticipated in previous chapters, this is already in use in technologically advanced countries, but it should find wider application.

In addition, the temporary solutions that have been carried out by the main airlines and shipping companies, such as the travels to nowhere, were nothing but makeshift ideas to recover economic loss, but they have certainly not been able to balance the losses. Moreover, these initiatives have often been accused of being unsustainable and not environmentally friendly. The trips have been described as unnecessary by several environmental associations around the world. Is this judgment correct? Civil aviation is considered to be an important factor in climate change. In fact, greenhouse gas emissions from aircraft have more than doubled in the last 20 years, mainly due to the increase in flights. Before the pandemic, the aviation industry alone emitted one billion tons of carbon dioxide into the atmosphere each year, including other types of emissions, such as nitrogen monoxide (NO) and nitrogen oxide (N_2O), often ignored but with effects up to three times higher than those related to CO_2 (e.g., Gössling and Higham, 2020; Gössling and Humpe, 2020; Koçak, Ulucak and Şentürk Ulucak, 2020; Prideaux, Thompson and Pabel, 2020; Sun, Lin and Higham, 2020). Aviation, as a human factor, accounts for almost 5 percent of global warming.

At the same time, as is known to all, ships are fuel-powered. According to a 2019 study by the Transport and Environment group, in 2017 the luxury cruise brands owned by Carnival Corporation & PLC emitted in European seas the volume of sulfur oxides from about 200 ships, and this volume was higher than that of all the cars in Europe combined (Transport and Environment, 2019).

However, as we anticipated, the new initiatives carried out during the pandemic period were not formulated only for economic purposes, but also to satisfy the need of people to resume normality by traveling. Their success, but also their environmental impact must represent the starting point for thinking about innovative and environmentally friendly modes of mobility.

The role of transport in sustainable development was already recognized at the 1992 United Nations Summit in Agenda 21 (Corbisiero and Minervini, 2017). The belief that transport and mobility can be fundamental levers for sustainable development has grown in recent years. Consequently, on the 2030 Agenda, sustainable transport has been included in several sustainable development goals (SDGs) and targets: SDG11 (sustainable cities), but also SDG3 (health and well-being) and SDG9 (business, innovation and infrastructure). Furthermore, the importance of the transport sector for the climate (SDG13) was further recalled in the United Nations Framework Convention on Climate Change (UNFCCC).

As a result, future mobility policies will have to promote sustainability by planning investments able to maximize their effectiveness and minimize economic and environmental costs. In this framework, the transport sector will be called upon to identify new models of tourism demand management capable of avoiding waste of resources or situations of congestion. The transition to the new normal cannot be characterized by a return to the "old mobility." For example, Renaud (2020) has suggested that a possible solution for the cruise tourism industry could be to adopt a "local mobility" model in which destinations ban large cruise ships but exert control over fleets of smaller ships.

Therefore, gradually the sector must be able to imagine unprecedented scenarios with a low environmental impact also thanks to the contribution of technological progress. This approach could allow the transport sector to occupy a strategic position to favor a resilient tourist restart that moves at the same time on social, geopolitical, green and technological levels.

Within this context, some initiatives are already taking shape. For example, the HyFlyer project, even on a small scale, launched the first commercial hydrogen aircraft on September 24, 2020. The M-class Piper refitted with an electric motor took off in Cranfield, England, marking a major milestone for the sustainable aviation sector. Its success bears the signature of the company ZeroAvia which, together with its partners European Marine

Energy Center (EMEC) and Intelligent Energy, has set itself the goal of decarbonizing small aircraft that make medium-distance journeys. The first flight of the hydrogen-powered commercial aircraft is just one example of the HyFlyer project. The final phase of the development program involved a journey of over 400 kilometers powered by fuel cells, taking off from an airport on the Orkney Islands, off the north of Scotland. The company's 2023 goal is a 19-passenger Twin Otter capable of 500-mile regional flights, while the long-term aim will target larger regional turboprops like Bombardier's Dash-8 or the ATR 500 series by the end of the decade.

A similar future is shared by maritime travel. The application of electric ships to reduce emission should be a direction of sustainable development. To cope with the difficulty of mobility, for example, the government of Singapore has decided to revolutionize sea transport, at least in its port. The Singapore Maritime and Port Authority has begun to convert all utility devices, such as local ferries, small cargo-unloading vessels, barges and the pilot vessels that drive larger vessels into electric devices. At the same time, it has also decided to set up the necessary recharging stations in port infrastructures.

Rail transport can be more environmentally friendly too, even though from an environmental impact point of view the train is already more energy efficient, as it is mechanical transport by land. An even greener solution could be the introduction of new, clean fuels with lower carbon emission than fossil fuels, such as Alstom's iLint hydrogen train, or prototypes running on liquefied natural gas being tested in Spain. Likewise, hydrogen batteries, as a source of electricity on board, only emit water, which is enjoying great media exposure and attracting large investments from the railway industry.

In order for the aforementioned good practices not to remain isolated cases, it appears necessary to prepare a clear regulatory framework and global guidelines that are able to direct the choices of the operators and customers of the transport system.

Bibliography

Almeida, C M and Costa, C M (2012) 'A operação das companhias aéreas de baixo custo na Europa. O caso da Ryanair', *Rivista Turismo & Desenvolvimento*, 1, 17–18: 387–402.

AAPA (2021) *Asia Pacific Airlines January Traffic Results*, Kuala Lampur: Association of Asia Pacific Airlines.

Arrigo, U and Giuricin, A (2006) *Gli Effetti della Liberalizzazione del Trasporto Aereo e il Ruolo delle Compagnie Low Cost. Un confronto USA-Europa*, Pavia: Società Italiana di Economia Pubblica.

ART (2020) *Indagine Sulla Mobilità dei Cittadini e Azioni di Spinta Gentile*, Turin: Autorità di Regolazione Trasporti.

Asgari, N (2020) 'Ryanair's bond draws €4bn in orders as Covid fears swirl', *Financial Times Global Economic Crisis*, 9/08/2020.

ATAG (2021) *The Impact of COVID-19 on Aviation*, Mexico City: Air Transport Action Group.

Blalock, G, Kadiyali, V and Simon, D H (2007) 'The impact of post-9/11 airport security measures on the demand for air travel', *The Journal of Law and Economics*, 50: 4.

Boros, L, Dudás, G and Kovalcsik, T (2020) 'The effects of COVID-19 on Airbnb', *Hungarian Geographical Bulletin*, 69, 4: 363–381.

Bruinsma, F, Pels, E, Rietveld, P, Priemus, H and van Wee, B (2008) *Railway Development Impacts on Urban Dynamics*, London: Springer.

Bucak, T and Yiğit, S (2021) 'The future of the chef occupation and the food and beverage sector after the COVID-19 outbreak: Opinions of Turkish chefs', *International Journal of Hospitality Management* 92: 102682.

Budd, L and Ison, S (2020) 'Responsible transport: A post-COVID agenda for transport policy and practice', *Transportation Research Interdisciplinary Perspectives*, 6: 100151.

Buehler, R and Pucher, J (2011) 'Making public transport financially sustainable', *Transport Policy*, 18, 1: 126–138.

Calle-Sánchez, J, Molina-García, M, Alonso, J I and Fernández-Durán, A (2013) 'Long term evolution in high speed railway environments: Feasibility and challenges', *Bell Labs Technical Journal*, 18, 2: 237–253.

CER (2021) *2020 Figures Reveal €26 Billion Loss for Railways Due to COVID-19*, Brussels: Community of European Railway and Infrastructure Companies.

Ciganik, L and Balasicova, I (2008) 'Protection and defence of railway transport against international terrorism', *Communications. Scientific Letters of the University of Zilina*, 10, 1: 40–44.

CIRIUM (2020) *Airline Insights Review 2020*, London: Cirium.

Corbisiero, F and Minervini, D (2017) 'Enviromental policies', in L L Lowry (ed.) *The SAGE International Encyclopedia of Travel and Tourism*, Thousand Oaks: Sage.

China State Railway Group (2020) *Annual Report of China State Railway Group*, Beijing: China State Railway Group.

China State Railway Group (2021) *Rail Transport Before Spring Festival Transport Reduced Passengers and Increased Load*, Beijing: China State Railway Group.

Choquet, A and Sam-Lefebvre, A (2021) 'Ports closed to cruise ships in the context of COVID-19: What choices are there for coastal states?', *Annals of Tourism Research*, 86: 103066.

Diaconu, L (2012) 'The evolution of the European low-cost airlines' business models. Ryanair case study', *Procedia. Social and Behavioral Sciences*, 62: 342–346.

Dobruszkes, F (2006) 'An analysis of European low-cost airlines and their networks', *Journal of Transport Geography*, 14, 4: 249–264.

Dobruszkes, F (2013) 'The geography of European low-cost airline networks: A contemporary analysis', *Journal of Transport Geography*, 28: 75–88.

Dube, K, Nhamo, G and Chikodzi, D (2020) 'COVID-19 cripples global restaurant and hospitality industry', *Current Issues in Tourism*, 24, 11: 1487–1490.

European Commission (2020) *Guidelines on Protection of Health, Repatriation and Travel Arrangements for Seafarers, Passengers and Other Persons on Board Ships*, Brussels: European Commission.

Fox, S (2014) 'Safety and security: The influence of 9/11 to the EU framework for air carriers and aircraft operators', *Research in Transportation Economics*, 45: 24–33.

Francis, G, Humphreys, I, Ison, S and Aicken, M (2006) 'Where next for low cost airlines? A spatial and temporal comparative study', *Journal of Transport Geography*, 14, 2: 83–94.

Gössling, S and Higham, J (2020) 'The low-carbon imperative: Destination management under urgent climate change', *Journal of Travel Research*, 60, 6: 1167–1179.

Gössling, S and Humpe, A (2020) 'The global scale, distribution and growth of aviation: Implications for climate change', *Global Environmental Change*, 65: 102194.

Hallion, R (2003) *Taking Flight: Inventing the Aerial Age, from Antiquity Through the First World War*, New York: Oxford University Press.

Ho, S, Xing, W, Wu, W and Lee, C (2020) 'The impact of COVID-19 on freight transport: Evidence from China', *MethodsX*, 8: 101200.

IATA (2020a) *COVID-19. Outlook for Air Transport and the Airline Industry*, Montréal: International Air Transport Association.

IATA (2020b) *IATA COVID-19 Survey*, Montréal: International Air Transport Association.

IdeaWorks (2020) *Flight Plan 2020 Report*, London: IdeaWorks.

IMO (2020) *International Maritime Organization: Coronavirus (COVID-19)—Preliminary list of recommendations for Governments and relevant national authorities on the facilitation of maritime trade during the COVID-19 pandemic, Circular Letter No.4204/Add.6*, 03/27/2020.

Jakab, P L (1990) *Visions of a Flying Machine: The Wright Brothers and the Process of Invention*, Washington: Smithsonian Books.

Jones, M (2012) *Lancashire Railways. The History of Steam*, Newbury: Countryside Books.

JRC (2020) *Future of Transport: Update on the Economic Impacts of COVID-19*, Ispra: Joint Research Center.

Kim, J, Kim, J and Wang, Y (2021) 'Uncertainty risks and strategic reaction of restaurant firms amid COVID-19: Evidence from China', *International Journal of Hospitality Management*, 92: 102752.

Kobayashi, K and Okumura, M (1997) 'The growth of city systems with high-speed railway systems', *Annals of Regional Science*, 31: 39–56.

Koçak, E, Ulucak, R and Şentürk Ulucak, Z (2020) 'The impact of tourism developments on CO_2 emissions: An advanced panel data estimation', *Tourism Management Perspectives*, 33: 100611.

Lieberman, C A and Bucht, R (2009) 'Rail transport security', in M Haberfeld and A Hassell (eds.) *A New Understanding of Terrorism*, New York: Springer.

Löfgren, O (1999) *On Holiday a History of Vacationing*, Berkeley: University of California Press.

Lyon, D (2006) 'Airport screening, surveillance, and social sorting: Canadian responses to 9/11 in context', *Canadian Journal of Criminology and Criminal Justice*, 48, 3: 397–411.

Malighetti, P, Paleari, S and Redondi, R (2009) 'Pricing strategies of low-cost airlines: The Ryanair case study', *Journal of Air Transport Management*, 15, 4: 195–203.

Moloney, K and Moloney, S (2020) 'Australian quarantine policy: From centralization to coordination with mid-pandemic COVID-19 shifts', *Public Administration Review*, 80, 4: 671–682.

Mulley, C and Nelson, J D (2009) 'Flexible transport services: A new market opportunity for public transport', *Research in Transportation Economics*, 25, 1: 39–45.

Nakazawa, E, Ino, H and Akabayashi, A (2020) 'Chronology of COVID-19 cases on the diamond princess cruise ship and ethical considerations: A report from Japan', *Disaster Medicine and Public Health Preparedness*, 14, 4: 506–513.

OECD (2020) *Tourism Trends and Policies*, Paris: Organization for Economic Cooperation and Development.

Paniccia, I (2020) 'Substitution or integration between traditional public transport and new forms of mobility. Implications for economic regulation', *Network Industries Quarterly*, 22, 3.

Pels, E (2008) 'Airline network competition: Full-service airlines, low-cost airlines and long-haul markets', *Research in Transportation Economics*, 24, 1: 68–74.

Pels, E, Njegovan, N and Behrens, C (2009) 'Low-cost airlines and airport competition. Transportation research', *Logistics and Transportation Review*, 45, 2: 335–344.

Peterson, J and Treat, A (2008) 'The post-9/11 global framework for cargo security', *Journal of International Economic Law*, 1.

Pirie, G (2014) 'Tracking railway histories', *Journal of Transport History*, 35, 2: 242–248.

Plant, J F (2004) 'Terrorism and the railroads: Redefining security in the wake of 9/11', *Review of Policy Research*, 21, 3: 293–305.

Prideaux, B, Thompson, M and Pabel, A (2020) 'Lessons from COVID-19 can prepare global tourism for the economic transformation needed to combat climate change', Tourism Geographies, 22, 3: 667–678.

Priemus, H and Konings, R (2001) 'Light rail in urban regions: What Dutch policymakers could learn from experiences in France, Germany and Japan', *Journal of Transport Geography*, 9, 3: 187–198.

Quigley, A L, Nguyen, P Y, Stone, H, Lim, S and MacIntyre, C R (2020) 'Cruise ship travel and the spread of COVID-19: Australia as a case study', *International Journal of Travel Medicine and Global Health*, 9, 1: 10–18.

Renaud, L (2020) 'Reconsidering global mobility-distancing from mass cruise tourism in the aftermath of COVID-19', *Tourism Geographies*, 22, 3: 679–689.

Rocklöv, J, Sjödin, H and A Wilder-Smith, M D (2020) 'COVID-19 outbreak on the diamond princess cruise ship: Estimating the epidemic potential and effectiveness of public health countermeasures', *Journal of Travel Medicine*, 27, 3.

Roth, R and Jacolin, H (2013) *Eastern European Railways in Transition: Nineteenth to Twenty-First Centuries*, Oxon: Routledge.

Roy, H, Gupta, V, Faroque, V R and Patel, A (2021) 'The impact of COVID-19 on the foodservice industry in Vancouver, British Columbia, Canada', *Anatolia*, 32, 1: 157–160.

Siddhartha, J (2020) 'Effect of COVID-19 on restaurant industry. How to cope with changing demand', *SSRN Electronic Journal*, 3577764: 1–3.

Sklarewitz, N (1991) 'Cruise company handles crisis by the book', *Public Relations Journal*, 47: 34–36.

Song, H J, Yeon, J and Lee, S (2021) 'Impact of the COVID-19 pandemic: Evidence from the U.S. restaurant industry', *International Journal of Hospitality Management*, 92: 102702.

Stevenson, D (1999) 'War by timetable? The railway race before 1914', *Past & Present*, 162: 163–194.

Suau-Sanchez, P, Voltes-Dorta, A and Cugueró-Escofet, N (2020) 'An early assessment of the impact of COVID-19 on air transport: Just another crisis or the end of aviation as we know it?', *Journal of Transport Geography*, 86: 102749.

Sun, Y, Lin, P and Higham, J (2020) 'Managing tourism emissions through optimizing the tourism demand mix: Concept and analysis', *Tourism Management* 81: 104161.

Swadi, T, Geoghegan, J L, Devine, T, McElnay, C, Sherwood, J, Shoemack, P and de Ligt, J (2021) 'Genomic evidence of in-flight transmission of SARS-CoV-2 despite predeparture testing', *Emerging Infectious Diseases*, 27, 3: 687–693.

Teberga de Paula, A and Merlotti Herédia, V B (2020) 'COVID-19 and cruise ships: A drama announced', *Études Caribéennes*, 47.

Transport and Environment (2019) *One Corporation to Pollute Them All Luxury Cruise Air Emissions in Europe*, Brussels: European Federation for Transport and Environment AISBL.

World Economic Forum (2020) *5G Outlook Series: Transforming Essential Services for Economic Recovery in the Great Reset*, Geneva: World Economic Forum.

Yamahata, Y and Shibata, A (2020) 'Preparation for quarantine on the cruise ship diamond princess in Japan due to COVID-19', *JMIR Public Health and Surveillance*, 6, 2: e18821.

Zhang, J, Hayashi, Y and Frank, L D (2021) 'COVID-19 and transport: Findings from a world-wide expert survey', *Transport Policy*, 103: 68–85.

Zhang, X and Wang, C (2021) 'Prevention and control of COVID-19 pandemic on international cruise ships: The legal controversies', *Healthcare*, 9, 3: 281.

5 Being a tourist without moving

Stationary tourism as an alternative strategy for traveling

Introduction

As Manovich (2013) noted, life beyond the physical world is occupying more and more space in contemporary society. As a result, over time many traditional activities have also been transformed into online realization. For example, online purchases and bookings are increasingly widespread and have thus encouraged disintermediation.

The emergency of the COVID-19 pandemic has further encouraged the use the Internet to connect people to the outside world. For example, restrictions imposed by the pandemic affected the entire school system, from administrative staff to teachers, students and their families. Most classrooms have shifted to remote learning since spring 2020, and many schools have also implemented a hybrid learning system thereafter. Therefore, an enormous effort has been made to ensure the continuation of teaching schedules, implementing strategies to redefine distance learning activities (e.g., Carpenter and Dunn, 2020; Yates *et al.*, 2020; Kenway and Epstein, 2021; Trinidad, 2021).

Similarly, the enforced home quarantine against COVID-19 led to the relocation of various public and private work activities to the home. Working from home has involved a large number of workers and professionals, stimulating a rethinking of work in an unprecedented scenario that takes into account production contexts and technological development (e.g., Abulibdeh, 2020; Matthewman and Huppatz, 2020; Žižek, 2020; Jenkins and Smith, 2021).

At same time, unexpected scenarios of tourism renovation have occurred in a very short time. The previous pages of this book present a detailed picture of the worst crisis that global tourism has experienced, but also the solutions that have been given at different levels to tide the sector over the difficulties caused by the pandemic. Many of the most innovative solutions exerted desirable effects thanks to the use of new technologies, providing effective support at the moment of standstill.

Compared with other sectors, technology has accentuated even more the double significance of tourism. In particular, during periods of travel bans,

DOI: 10.4324/9781003195177-6

alternative digital travel practices have become increasingly common. In this scenario, the digital reproduction of reality and the birth of new intangible spaces for tourist practices have occupied a central space, since they have made it clear that the exploration of the world can also take place in a confined space, in which travel and motion are partially or completely banned.

In order to better understand the complexity of the tourist experience and the way in which tourism-related activities can be carried out, it is useful to distinguish between "moving tourism" and "stationary tourism." The former represents tourism in its traditional sense. It depends on physical mobility, travel by means of transport, and in-person encounters between tourists and local communities. In other words, "moving tourism" presupposes the movement of people who personally visit tourist places, instead of knowing them through the use technologies. It can only come about when reality is available.

I decided to name traditional tourism as "moving tourism" for two main reasons. First, because its fulcrum is represented precisely by the physical interaction between tourists and destinations inasmuch as traveling takes place in a territory other than that of departure. That is to say, moving tourism surpasses the physical distance that separates the traveler's place of residence and the tourist destination. Moving tourism, however, can be further categorized into domestic tourism and cross-national (e.g., international or intercontinental) tourism.

Second, I chose to use the expression "moving" in the sense of emotional transport too. In fact, moving tourism requires the involvement of all five senses during the travel experience. Thus, tourists who reach a destination see with their own eyes the natural and cultural beauties of the place; they can touch the artifacts and interact with people; they smell the typical smells of the place; they taste the local cuisine; they listen with their own ears to typical sounds, voices and noises, but also national languages, regional dialects or typical local expressions.

In this type of experience, technologies are present, but they are not indispensable. In other words, during a moving tourism experience technologies can be considered as tools to support visitors. As the literature on the subject has pointed out, during travel experiences more and more tourists rely on new technologies to communicate, read and leave reviews online, make reservations and so on (e.g., Buhalis and Licata, 2002; Main, 2002; Soteriades, Aivalis and Varvaressos, 2004; Zhou, 2004; Cox *et al.*, 2009; Gruescu, Nanu and Pirvu, 2009; Berné *et al.*, 2015).

As we know, the pandemic has often prevented travelers from the experience of travel, imposing restrictions on mobility, thus affecting distances among people. As a result, an alternative means of connecting tourists to destinations has developed, that I put under the umbrella term "stationary tourism." I borrowed the adjective stationary from the world of gymnastics, thinking about the metaphor of "stationary bicycles." As it is known, they are exercise equipment placed in a fixed place (at home, in the gym or in other indoor or outdoor places) that allow people to simulate the use of a bicycle for physical training as if they were riding a traditional bike. This type of device is very suggestive because it allows people to move even if they actually remain still in the same physical point. Thus, just as a stationary bicycle remains in place but allows people to pedal at different speeds and intensities, offering the same sensation as a regular bike, a stationary tourist is able to make a journey without changing location through various possible intangible forms of travel, such as virtual tours, online gaming and remote online events. In this sense, stationary tourism is not the opposite of moving tourism, but it represents a possible substitute travel experience that transforms tourism into a kind of performance, freeing itself from the constraints of space and time. It is clear in these kinds of experiences, the relationship among tourists and territories is formed through media, thanks to technological tools, which let tourists get a sense of places and attractions in order to explore destinations without actually being there.

First of all, one of the main characteristics of this kind of tourism is its temporality. While moving tourism begins with a one-way trip and ends with a return trip, stationary tourism begins and ends in the same place in a well-defined period of time, delimited by the use of technologies. Stationary travelers find themselves catapulted from one moment to the next into this kind of experience, immediately benefiting from the immaterial journey.

Second, stationary tourism is a social experience that implies the cognitive rather than physical involvement of tourists. In other words, travelers have no physical experience with places, people and objects. On the contrary, the tourist activity is charged with new forms and symbolic modalities that require a mental involvement.

Finally, depending on the type of experience that tourists have, they are able to involve one or more of the senses, but not all their senses can be activated at the same time. Certainly, in this form of tourism, the sense of sight is the greatest protagonist (for the enjoyment of photographic or textual material), followed on many occasions by hearing (audio, video and music). Touch is only rarely stimulated, except on specific occasions when particular sensors are used. At the moment, in most cases taste and smell are still excluded from stationary tourism. People who don't practice moving tourism are aware that they have to give up this kind of sensation.

We know that sociology has long pointed out that one of the main characteristics of contemporary society is the presence of reticular links (Castells, 1996) that have become increasingly mobile also through the use of communication technologies that allow people to establish and maintain relationships at a distance. So, as Urry suggests "These convergences of travel and communications further transform the character of co-presence that is increasingly mobilized" (Urry, 2002: 24). In other words, the new technological devices, through the network connection, have given rise to a multiplication of interactions out of the production and reproduction of spaces.

Unlike in the past, these kinds of trips could be considered as a useful way for traveling with greater safety and more opportunities. In this sense, the "stationary tourism" in the pandemic scenario can be defined as a "mobile immobility," which allows tourists to reach the destinations they intended to visit without actually being in that place, while maintaining social relationships and carrying out tourism activities through media.

In other words, "stationary tourism" represents a practical tool for understanding the intrinsically possible alternatives to "moving tourism" which in any case constitutes a socially significant experience. Indeed, it has diverted the attention of many tourists into virtual manifestations of places such as streets, squares, museums and archaeological parks, overcoming the limited carrying capacity as well as the spatial and social constraints among places of residence and tourist destinations.

Online photos or videos, Google Maps, Google Street View, three-dimensional (3D) reproductions and videogames, and virtual tourism initiatives are all tools that together contribute to building and making "stationary tourism" feasible. This chapter reviews the main "stationary tourism" offers from all over the world emerged during the pandemic period.

5.1 Tourist complexity redesigned by new technologies

Information technology has contributed enormously to providing people with the opportunity to experience the world and practice tourism without actually visiting a place. In this regard, the digital representation of places to visit has been made possible by the diffusion of intangible environments in alternative formats. These are a visually realistic reproduction of 3D completely computer-generated environments, which offer people the possibility of visiting places, sometimes even going back to the past.

Their main feature is represented by the fact that they make users feel like they are immersing themselves in the environment represented (e.g., Andreoli, 2018; Wagler and Hanus, 2018; Mohanty, Hassan and Ekis, 2020).

Virtual reality is not new, even though the pandemic has certainly contributed to accelerating its spread globally, especially its application for experiencing "stationary tourism." As early as in the 1950s, the American director Morton Leonard Heilig experimentally created equipment that gave viewers multisensory experiences while watching movies (Steinicke, 2016). The audience not only experienced these in films but through the so-called "Sensorama" all their senses were stimulated. This equipment provided the viewers with visual, auditory and movement stimuli. In fact, the "Sensorama" was also able to show stereoscopic images, spread aromas and perfumes, and make sounds, emanate into the air to simulate the effect of the wind and reproduce tactile sensations thanks to the use of a handlebar.

In the early 1960s, Heilig also designed a portable version of the "Sensorama," which was called a "Head Mounted Display." This was a viewer that was equipped with lenses with 140-degree horizontal and vertical viewing angles that also included stereo headphones and an air vent.

Even if the excessive cost of the equipment and the lack of funding forced the large US movie production companies to put an end to the experimentation, Heilig can be defined in all respects as a vanguard in the construction of a reality beyond the physical world. His prototype is certainly different from the contemporary format which allows people to do "stationary tourism," but it represents the first step that was ever taken in the IT field to allow people to gain a mediated experience, reaching a distant place while sitting in a room.

Toward the end of the 1960s, the computer scientist Ivan Sutherland built the first real computer system capable of showing 3D images projected on real objects (Earnshaw, Gigante and Jones, 1993). His invention represents the prototype of augmented reality in history, which has since been perfected and made more accessible with the passage of time. In particular, Sutherland's invention made a 360-degree view of 3D environments available, which, thanks to the presence of a motion sensor, changed according to the user's position. The equipment was fixed to the ceiling and connected to the user's head through a long mechanical arm that supported its weight. For this reason, it was nicknamed "The Sword of Damocles." Sutherland also patented virtual reality glasses, a technology that became very popular in the 1980s. In addition, in 1988 he won the Turing Prize for his invention of the Sketchpad software, which is the predecessor of the most widely used interfaces in computer graphics.

In 1987 the term "virtual reality" began to enter the common language thanks to Jaron Lanier, who created the first 3D commercial software (e.g., Heeter, 1992; Cruz-Neira, Sandin and De Fanti, 1993).

The term "augmented reality," in fact, was not coined until many years later, in 1992, when researchers Thomas Preston Caudell and David Mizell were commissioned by the Boeing company to find a solution to speed up the technician's task of wiring electrical systems in confined aircraft fuselages (Sudharshan, 2020). Caudell and Mizell had the intuition to equip technicians with a wearable computer, a mini head-mounted computer which projected wiring diagrams on special viewers, so that the technicians could consult the superimposed diagrams on the parts of the aircraft on which they were working. Later this technology was applied to fighter aircraft to aid flying and training: through a display mounted on the glass of the aircraft cockpit, the pilot could directly view flight data (e.g., altitude, speed, inclination of the aircraft, distance from the target) while keeping his gaze fixed on what was in front of him.

Augmented reality can be defined as a technology based on the principle of superposition. In other words, using appropriate viewers, such as a display, a pair of glasses or the screen of portable devices, it is able to add information and multimedia content to the surrounding reality (Furht, 2011). To be more specific, it helps make users feel as if the virtual objects

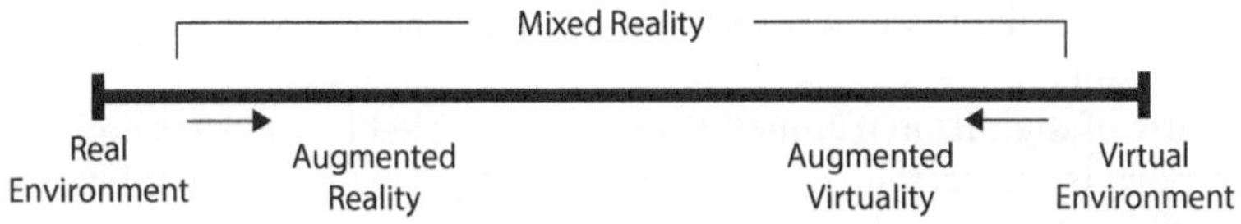

Figure 5.1 Reality–virtuality continuum.

Source: Author's elaboration of Milgram and Kishino (1994).

are really present and integrated in the real world and see these objects move from different angles. The devices recognize framed objects and activate a series of information and contents such as text, video, audio, 3D objects, giving the user the feeling that virtual and real objects coexist in the same space and that s/he can move and interact with them freely. The real world is "augmented" or virtually enriched by additional synchronized computer-generated visual, graphic and textual information, which would not be detected directly by one's senses (De Paolis, 2012).

Obviously, even if augmented reality and virtual reality both contribute to enriching the real world, they are fundamentally different. According to the theories of Milgram and Kishino (1994), these are at the extremes of a continuum (see Fig. 5.1): virtual reality is on the far right and can sometimes isolate people from the real world, in a sort of awake hallucination. In fact, it can be totally immersive when the user is completely isolated from the real environment and is transferred into a parallel 3D reality generated by a computer through the wearing of special technological devices (such as a 3D viewer in the form of glasses or a helmet, a pair of earphones, gloves or even a cyber-suit) (Azuna, 1997; Gutierrez, Vexo and Thalmann, 2008). Nonimmersive virtual reality allows people to experience an "augmented virtuality," which is in some respects less engaging, as the subjects accesses the virtual spaces in which they can walk and visit, without isolating her/himself from the starting context (e.g., Carmigniani *et al.*, 2011; Meschini, 2011). This type of experience involves a combination of physical environment and virtuality that is realized not only in immateriality (Montagna, 2018).

Within the reality–virtuality continuum, augmented reality lies on the left of the scale (Fig. 5.1), closer to the real environment, whereby the real world predominates over the data added by computer. The intermediate space between an absolutely immersive virtual reality and a technology-free environmental reality has been defined as mixed reality (Mandelli and Accosto, 2012).

In more recent times, Internet 2.0 has been one of the main elements producing a further mechanism of social transformation in a significant manner, so much so that we could speak of a "2.0 society" in computer language (e.g., Meyrowitz, 1985; Rheingold, 1991; van Dijk, 1999, Meyrowitz, 2005; Granieri, 2006; Monaco, 2015). In particular, the use of Web 2.0 has made the progressive digitization of everyday objects and all the territories in the world possible.

The use of digital environments has been enabled by websites, video and audio portals, devices such as viewers, smart glasses and apps, but above all by smartphones, which are ideal carriers thanks to the fact they can be imbedded with digital cameras, GPS tracking systems, compasses and Internet connections (Hwang and Yu, 2012; Lee, Tewolde and Kwon, 2014; Lew, Hall and Williams, 2014).

Of all the platforms that show the streets of major cities around the globe in 3D, the best known is GoogleMaps, and many other apps on smartphones allow people to travel around the world without leaving their place of residence. Orbulus is one of the most downloaded of these apps, offering a 360-degree perspective on the main monuments and tourist attractions of the world, presenting a gallery of images created by Google Photo Sphere. Similarly, the popularity of video games has also contributed to spreading this type of experience. Since modern consoles have entered people's homes more and more massively, many games have been made with real settings as a reference (e.g., Consalvo, 2006; Umaschi Bers, 2010; Woods, 2012; Baer and Hui, 2019). The main elements underlying the success of video games in promoting tourism have been, on the one hand, the graphic characteristics that have improved a lot over time and, and on the other, the integration of online and social components with these games (starting about 10 years ago). Connecting players from all over the world was an action that led to the huge expansion of video games, making it possible to have mediated contact between travelers and local communities.

In this regard, online game-streaming platforms have recorded an increase in the numbers of audience and content since 2020. During home quarantine due to the pandemic, live steaming has consolidated its role as provider of entertainment, remote social interaction and world news updates. Quantitatively on a global level, in 2020 both Twitch and Facebook Gaming have recorded an increase of 80 percent in audience compared to in 2019 (StreamElements, 2021). These platforms can be considered a compound of TV and social networks. On them game anchors broadcast live and comment on games and performances together with their spectators. When it comes to the integration of virtuality and tourism, several scholars questioned the possible implications: some have been quite critical (e.g., Mandich, 1996; Refsland *et al.*, 2000; Bonesio, 2002; Jansson, 2002; Jacobson and Holden, 2005; Koltko-Rivera, 2005; Brunelli, 2017), because they believe that virtual environments devalue the essence of social reality, offering a reflected image of the world. The idea behind these arguments is that the digital reproduction of territories and tourist attractions weakens the symbolic significance of tourist places, including memory, identity and beauty, creating indistinct virtual environments. In other words, the main fear is that virtual representation would have the demerit of altering people's perception of the material world.

This type of debate cannot fail to be reminiscent of the theses written by Walter Benjamin. In 1936, in his work entitled *Das Kunstwerkim Zeitalter seiner technischen Reproduzierbarkeit* (The Work of Art in the Age of Mechanical Reproduction), he formulated an idea: like all other consumer goods, art was gradually becoming subject to mechanical reproducibility (Benjamin, 1936). At the same time, the new technologies were appearing on the scene.

First with photography and later with cinema, Benjamin underlined how a process of quantitative acceleration of the cultural heritage reproduction was beginning to spread in the first half of the twentieth century. This quantitative transformation had a qualitative nature concerning the products of human ingenuity.

According to Benjamin, works of art were entering a new era, with the loss of their aura, or to be specific, their charm.

As reported by the most modern detractors of digital reproductions, the virtual representation of territories can pose a threat instead of a resource because it induces people into a fictitious experience, which cannot be considered significant. In this regard, a few years ago the French sociologist Baudrillard wrote "An image is an abstraction of the world in two dimensions. It takes away a dimension from the real world, and by this very fact the image inaugurates the power of illusion. On the other hand, virtuality, by making us enter into the image, by recreating a realistic image in three dimensions (and even in adding a sort of fourth dimension to the real, so as to make it in some way hyperreal), destroys this illusion [...]. Virtuality tends toward the perfect illusion. But it isn't the same creative illusion as that of the image. It is a 'recreating' illusion, revivalist, realistic, mimetic, hologrammatic. It abolishes the game of illusion by the perfection of the reproduction, in the virtual rendition of the real. And so we witness the extermination of the real by its double" (Baudrillard, 1997: 9).

In other words, the tourist heritage in its digitized version cannot be considered an element of value, because it lacks its real essence (e.g., Thomas and Mintz, 1998; Jalla, 2019).

In addition, some scholars (e.g., Cheong, 1995; Guttentag, 2010) have pointed out that remote tourism experiences could exert negative impacts on the economy of the real destination, as people would spend money on technologies to visit places from afar, while local communities would face a significant loss in customer flow and revenue.

Finally, the philosopher Erick Ramirez recently argued in an interview (Tanuvi, 2020): "Virtual reality simulations have to be created by people for people. Like most technologies, this means that the people responsible for creating virtual travel packages have a lot of control over how vacation destinations are represented to virtual tourists. One of the important benefits of tourism is that the tourist is usually forced to engage with native populations on their own terms (cruise ships and tropical all-inclusive resorts may be the exemption to the rule). This means that tourists can often learn that their preconceived notions of what other peoples or nations are like were wrong. With VR travel, this isn't possible."

However, other studies (e.g., Williams and Hobson, 1995; Burdea and Coiffet, 2003) emphasize that virtuality does not replace reality or subtract anything from it. On the contrary, digitalization enriches the physical world as well as innovating the tourism market. In general, all the technologies that we have briefly described have obviously been able to positively affect the perceptions of their objects, contributing to the experience in terms of sensitivity and exploration of the territories.

Under these circumstances, tourism is experiencing a new season in terms of the new forms through which mobility is achieved. As Manovich (2001) suggests, if we accept this spatial metaphor, the nineteenth-century European flâneurs and flâneuses find their reincarnation in the figure of the digital travelers. More specifically, throughout the history of urban civilization, engineers, architects, sociologists, artists and urban planners have designed and built roads, palaces, urban structures of great artistic and cultural value, which are re-proposed as intangible places for stationary flâneuristic walks. Thus, the possibility of discovering tourist heritage by crossing digital territories evokes the slowness of strolling and the rituality that are typical of flânerie that takes place through curious and solitary walks (Gros, 2009; Nuvolati, 2013).

As we will see in the next paragraphs, in times of pandemic, "stationary tourism" have offered a series of unprecedented possibilities, which are becoming very useful tools both for the recovery of tourism and meeting a tourism demand that otherwise would have been undersatisfied.

5.2 Some global initiatives

In recent years, many tourist attractions such as museums, archaeological sites, ancient architectures and historic relics have already been digitized in the form of 3D or virtual models, and many of them have been made available online. As we anticipated, this process accelerated sharply during the pandemic period. Thus, in order to enhance the construction of the world cultural heritage in an alternative way and at the same time, to encourage other tourists to practice a form of nonmobile tourism, many institutions and tourist sites have made their digital contents accessible for free or have organized themselves to allow record numbers of stationary tourism.

To begin with, since the early days of March 2020, the United Nations Educational, Scientific and Cultural Organization has been offering the general public free access to the World Digital Library. Founded in 2009, this is an international historical library under the management of UNESCO and the Library of Congress in collaboration with libraries, archives, museums, educational institutions and international organizations from 193 countries. The library collection has about 20,000 libraries from all over the world and from every era, written in Arabic, Chinese, French, English, Portuguese, Russian and Spanish. In particular, the materials available are in the form of books, interactive maps, photographs, magazines, newspapers, films, timelines and sound recordings. They can be dated back to as early as 8000 BC. Users can retrieve the information about a book by place, time, topic, type of element, language and contributing institution. Each article in the World Digital Library is accompanied by an explanatory description and videos offered by the curators.

Some similar initiatives have also been set up in individual territories of the world.

For example, in late January 2020, the National Cultural Heritage Administration (NCHA) in China strongly encouraged the country's museums to offer digital solutions for homebound citizens since that was the time of the Chinese New Year. In a short time, the already-existing "Panoramic Palace Museum" program, accessible on WeChat, was updated by most of the Chinese cultural institutions that had digitized their collections. From the Karamay Museum in the Xinjiang Uygur Autonomous Region to the Revolutionary History Museum in Shanghai city, 100 Chinese institutions responded promptly to the NCHA initiative by making their collections publicly available through online exhibitions via a centralized portal.

The two main models proposed by Chinese museums were virtual tours and detailed photo exhibitions. Virtual tours offered a panoramic view of the museums. Stay-at-home users could read descriptive cards of the museum works and zoom in on artifacts for a better understanding of, for example, the history of wrestling and competitive fishing at the Chinese Sports Heritage Exhibition in Zhengzhou.

As for the photo exhibits, the NCHA provided 3D items, giving visitors the chance to turn and inspect the valuable objects on their displays such as Neolithic pottery at the Shaanxi History Museum or an East Han jade screen in the Hebei Provincial Museum.

In addition to national cultural institutions, a number of independent arts institutions also actively responded to the call made by the NCHA. For example, the M Woods Museum of contemporary art in Beijing launched an experimental online exhibition on the platform that involved the participation of visitors in which they could play and communicate with each other through an interactive board.

In Singapore, Airbnb has started a collaboration with their Tourist Board to organize paid Virtual Trips through which tourists from all over the world have had the opportunity to admire the beauties of the city, take part in various cultural and social initiatives, as well as gain access to virtual tours of exhibitions and museums.

The National Museum of Modern and Contemporary Art in Seoul has also been accessible online thanks to Google's virtual tours. Since its inauguration in 1969, the museum has become a milestone of Korean art history.

In the periods after the arrival of the virus in Europe, the European Commission supported the activities of the "Europeana" digital cultural platform, which had already been active for some years, quickly enriching the site with new contents. It is a veritable multilingual online showcase of the cultural heritage of each member state and is designed to promote the digital use of cultural heritage. In particular, the cultural platform has a special section called "Discovering Europe" which contains a vast collection of materials in digital format (including films, photographs, works of art, sounds, new and old maps, manuscripts, books and newspapers) that come from museums, libraries, archives and audiovisual collections throughout Europe. The tools on Europeana allow users to make comparisons between different countries on the basis of their artistic heritage. The tourist experience made possible by the platform enables visitors not only access to the current cultural heritage of distant territories, but also contact with the past, thanks to the digitization of artifacts, images, videos and writings concerning the periods of the Roman Empire or the Renaissance.

Museums in various European countries have also taken steps to create alternative approaches to traditional visits, offering audiences around the globe virtual experiences. In Italy, for example, online visitors from all over the world have been able to admire the masterpieces of modern art exhibited inside the Pinacoteca Brera in Milan. In Florence, the Uffizi Gallery has prepared an experiential journey of the works from Botticelli to Cimabue in high definition. Another important example of Italian technological innovation designed to enhance the national historical-artistic heritage is certainly the Egyptian Museum of Turin, which has proposed interactive itineraries with 3D films and detailed reconstructions of the various phases of Egyptian history.

South Tyrol in Italy has also offered virtual tours in museums, the most interesting of which is certainly represented by the virtual visit to the Archaeological Museum of Bozen/Bolzano, where the mummy of Ötzi, "the man who came from the ice" is kept. Also noteworthy has been the #altoadigelooksforward initiative in the same area. It is a "virtual forum" which aims to bring together tourism operators to collect testimonies and ideas for the future.

In France, the Musée du Louvre has begun to offer virtual tours that allow stay-at-home people to see not only the many works it contains, but also to visit the famous glass pyramid of the building.

In the United Kingdom, the British Museum in London, and in Spain, the Prado in Madrid have made their works available online, classifying them by themes and eras. The Department of Tourism of the City Council of the Spanish capital has also published a series of virtual tours on the social channels of Visit Madrid where users have been able to visit the cultural sites and enjoy the best views of the city and green spaces from a distance.

In the United States, the Google Art Project has collaborated with institutions to offer the opportunity to visit the Metropolitan Museum in

New York, whose entire collection has been digitized into a real multimedia gallery where people could visit and examine not only works of art, but also photographs. In addition, Google's Street View has offered the opportunity to see, in a digital manner, the famous spiral staircase of the Solomon Guggenheim Museum in New York and to visit the National Gallery of Art in Washington, D.C. The visit to this famous American art museum could be accomplished through video tours, insights into the highlights of the collection, and audio and video recordings of lectures by artists and curators.

Thanks to the agreements of many museums in the world with Google Arts & Culture, Street View technology has allowed people admire works of art online, thus tourists from all over the world have had the opportunity to visit the J. Paul Getty Museum of Los Angeles which houses thousands of works of art from the periods ranging from the eighth to the twenty-first century. By virtually wandering through the rooms, visitors have had the opportunity to admire a vast collection of European, Asian and American paintings, drawings, sculptures, manuscripts, miniatures and photographs.

Under the same agreement, the National Museum of Anthropology in Mexico City has made its heritage digitally accessible. Without a doubt, this is the most important museum in the country. Built in 1964, it is entirely dedicated to the archeology and history of the country's pre-Hispanic heritage. On the online platform, visitors can access ancient artifacts, some of which can be dated back to the period of Mayan civilization.

In Australia, the New South Wales Government has created a specific page on its institutional website named "Virtual Tours from the comfort of your home" in which it lists the museums and galleries that offer virtual tours. A specific site for visiting the beautiful gardens and historic state rooms of Government House in Sydney has also been created. Furthermore, for the first time, thanks to an agreement with Google, the Sydney Biennale has allowed the public to interact in real time with artists through the website. The site also has implemented the function of live streaming, as well as podcasts and guided tours.

The Cape Town tourism authority in Africa has also launched a campaign to promote online tourism called "We Are Worth Waiting For." Thanks to collaboration with the Virtual Reality Company, it has offered many tours to visit the city virtually, including tours to Robben Island, with its former prison, and to Table Mountain.

According to UNWTO (2020) data, in early June 2020, 74 percent of African governments restricted tourists from entering their countries. Before the pandemic broke out, Africa was the fastest-growing tourist region. In 2018, approximately 67 million tourists visited the continent, generating $38 billion in revenue. In 2019, according to preliminary data, the number of tourists increased by 4.2 percent. The organization had estimated that Africa could have counted on a 3–4 percent increase in 2020 if it hadn't been for COVID-19.

5.3 Not just museums…

While visits to museums and the exploration of art collections have been plentiful, they have not been the only digital tourist experiences that have occurred in the world. Through a series of social campaigns and private schemes, the stationary tourism set up during the pandemic managed to satisfy other types of tourists as well. For example, for movie buffs, some countries have offered virtual tours to visit the shooting locations of the most famous movies, turning them into online destinations.

Traditionally, the so-called movie-induced tourism (Hudson and Ritchie, 2006) is able to attract different types of visitors, who want to combine the visit of the places with their passion for cinema. More specifically, sociological studies on the subject segmented movie fans into three macro categories depending on their level of involvement (Macionis, 2004): "general," which are people who do not visit a place just because it has been exposed in movie or drama, but appreciate it also for this reason; "serendipitous," namely people who are fascinated by the idea of being in the places where a film was shot, without taking part in activities related to this aspect; "specific," which are the real cine-tourists. For them, watching a film is both a push factor (since the location of the film is able to arouse intimate emotions) and a pull factor (because the vision of the film has a very powerful impact on destination image and the decision to travel) (Riley and Van Doren, 1992).

In the case of stationary movie-induced tourism, evidently it mainly affects the subjects that fall into the last category, since they do not want to be mere spectators of an entertaining show, but they want have a direct experience of the destination featured on television, video or movie screen, also taking part in engaging initiatives.

Thus, for this tourist niche, for example, during the first Austrian lockdown, the city of Salzburg organized a series of online visits to discover the places where the film "The Sound of Music" was shot. Numerous locations included the Benedictine Abbey of Nonnberg (the monastery where Maria lived) and Leopoldskron Castle (interior of the Von Trapp family home). The project was very successful, so much so that shortly after it was replicated to allow visitors to visit the settings of the movie trilogy dedicated to Princess Sissi, whose residence was rebuilt in the Schlosshotel Fuschl near Fuschl am See in Austria.

Similarly, in Central America, the Visit Central America organization has proposed a cinematic journey through five films and TV series, from the third series of "La Casa de Papel" (Islas San Blas in Panama) to the film "Jurassic Park" (Isla del Coco in Costa Rica); from "Pirates of the Caribbean" (Samanà, Dominican Republic), through "Terminator—Dark Destiny" (Izabal beach in Guatemala), to "Un lugaren el Caribe" (Roatan island in Honduras).

Still in the field of experiential tourism, some darker initiatives have been produced too.

For example, on the occasion of Halloween, the online booking platform Tiqets proposed a remote visit to some attractive places, such as Dracula's castle in Transylvania, so travelers could walk the ghostly corridors that inspired the atmosphere of Francis Ford Coppola's film "Bram Stoker's Dracula." The other two experiences organized by Tiqets were a live show in the legendary Warwick Castle (England) and a visit to the Madame Tussauds wax museum in Berlin (Germany), including a mystery murder that users had to solve via chat.

In the field of experiential tourism, two important players, Airbnb and Amazon, distinguished themselves by offering tourists the opportunity to take part in special online events. In April 2020, Airbnb started a collaboration with its hosts, asking them to offer online experiences with high-quality standards. This started in Italy, since it was the first European country to experience the lockdown. After a short time the same scheme extended to other countries. The Airbnb offer did not include cultural visits or thematic tours. Instead, the catalog of possibilities was full of global experiences from the United States to South America, from the Far East to Africa. It included cooking lessons, physical training with Olympic athletes, meditation sessions held by a Buddhist monk from Osaka, and how to make gin cocktails. The initiatives were offered in English and other languages mastered by the hosts, at a cost ranging from 1 to 40 euros. They were also open to groups through the Zoom platform. For activities that required materials, such as cooking classes or the natural cosmetics class, Airbnb sent a list of things that people needed to get before the appointment started.

Amazon launched its Amazon Explore platform in September, 2020, a new service to allow its customers to book virtual but interactive experiences. Bookers could contact local experts via their computer in real time. Experiences centered on tours of faraway places, but also on creativity cultivation, manual skills acquisition and, in some cases, shopping in boutiques around the world.

In its first phase, the service was available only in the United States and membership was by invitation.

Many other operators, both public and private, also expanded their online offers during and after the lockdown. For example, the National Tourism Organization of Japan made a 360-degree virtual reality film to continue the promotion of the Olympics 2020.

Likewise, due to the COVID-19 emergency, the city of Dubai had to postpone its Expo 2020 event. Under these circumstances, the emirate's tourism office created a 360-degree tour of the city, offering inhabitants the opportunity to discover its futuristic architecture, such as the Burj Khalifa, which, at 828 meters, is the highest building in the world, but also the long beaches of fine sand and The Palm, the palm-shaped artificial island which stretches into the open sea.

Similarly, National Geographic has created a platform to upload a series of 360-degree videos of wildlife and diving sites that, when paired with an

Oculus Quest VR headset, can make users feel like they are swimming with humpback whales in Antarctica, or exploring the coral reefs of Indonesia.

At the same time, the Australian tourist board launched the tourism and territorial promotion campaign, in the innovative 8D audio technology. This incorporated six videos that accompany viewers on a sensory journey, immersing them in unique sights and sounds.

By the same token, in the German town of Herrenberg, the HLRS (High Performance Computing Center) together with a team of researchers from the Fraunhofer Institute, the Kommunikationsbüro Ulmer and the University of Stuttgart developed "spatial syntax." Thanks to the application of augmented reality, the aerospace sector, geographic information systems (GIS) and the contribution of members of public who provided important data through an app, tourists across the globe could explore the historic center of the German town at a distance. The virtual representation took place in real time. In other words, people who connected to the platform could see what was happening inside the town, moving within the urban space by setting different parameters such as traffic, crowding on the street, time interval and so on.

5.4 Tourism experiences through games

While cinemas and theaters around the world have been closed for a long time for the sake of stemming the spread of the virus, many people have turned to home entertainment during times of isolation. Among the tools that people have used to discover the world without moving into physical space we find "gaming experiences." Recent years have seen a proliferation of online games in different contexts. Thanks to the significant contribution of gamification and storytelling, tourists now have the opportunity to get in touch with different territories and their artistic and cultural heritage, simply by playing online (e.g., Molz, 2004; Fulco, 2006; Montola, Stenros and Waern, 2009). In this sense, in gaming experiences, stationary tourists are not only users, but they must also experience a series of situations aimed at winning something (Olszewski, Pałka and Turek, 2018). Consequently, the so-called "homo game" (Pecchinenda, 2010) is forced to relate to the resources of the territory in which it is immersed. The tourist environment represented is not just a mere background, but acquires the function of a destandardized relational space.

A study conducted by the research and consulting firm Simon Kucher & Partners (Jaeger, Zarb and David, 2020), pointed out that the gaming sector was one of the few niches that experienced growth during the first half of 2020. More specifically, the study involved 13,000 representative consumers around the world, identifying nearly 5,000 gamers and asking them a series of questions. The research highlighted that globally there was a 39 percent increase in the amount of monthly consumption on games in May and June 2020 compared with the pre-COVID-19 period. This trend was consistent across nearly all age groups. Furthermore, the interviewees indicated that their new interest in gaming was not temporary, but also in the future they will continue to use this media product.

Even before the pandemic, some academics had already shown interest in studying the relationship between tourism and gaming (e.g., Bulencer and Egger, 2015; Xu, Buhalis and Weber, 2017; Alčaković, Pavlović and Popesku, 2018; Skinner, Sarpong and White, 2018), arguing that the play component could also represent a useful tool for the promotion and sponsorship of different territories. In fact, they allow users to immerse themselves in a simulated world of travel.

Although it is difficult to single out the specific characteristics of tourist gaming precisely, since there are several technological solutions that allow people to simulate a travel experience by playing (e.g., Terlutter and Capella, 2013; Weber, 2013; Seaborn, 2015), in general we can surmise that this type of experience generates positive experiences in users, stimulating their involvement and interest in the places they discover through playing (Brown, 2017).

Many initiatives that combine the experience of play and access to artistic and cultural heritage have been carried out during the pandemic.

In Europe, for example, one of the first structures that experimented with this type of venture was the French Georges Pompidou National Center for Art and Culture in Paris, which developed Prisme7, the first video game that allows people to explore the world of art on a multisensory level. The game was made by Olivier Mauco of "Game in society" and Abdel Bounane of "Bright," with the support of the Ministère de l'Education. By playing Prisme7, people could discover the universe of modern and contemporary art and interact with various works of art. They were also able to build their own virtual collection. The protagonist of the game, represented by a luminous molecular entity, had to move in colored spaces, experimenting with the plastic and sensory characteristics of about 40 works among the most representative of the museum's collection.

In May 2020, the Italian company "Play the World" organized a virtual treasure hunt for teams that sought to explore treasures in 25 places around the world. Users had to solve puzzles, riddles and math tests in order to find 25 stages in the footsteps of less famous or not-so-famous Italians. Participants moved from Mongolia to New Zealand, from the Pacific islands to Turkey via Google Street View.

Also in Italy, a game set in the world of the opera house was created by the Teatro Regio di Parma. The name of the online game was "A life in Music." It was available for free download from the App Store and Google Play Store, and recorded 200,000 downloads in 9 months. In nine acts and nine Verdi interludes, the game portrays a friendship formed in the summer of 2008 in the places associated with Giuseppe Verdi between two young people who have different personalities, but a common intense passion for classical music.

Moreover, the National Archaeological Museum of Naples (MANN) was the first archaeological museum to produce a video game for audiences around the world. The name of the game, launched in March 2021, is "Father and Son." It has both in English and Italian versions and allows people to explore the streets of the Neapolitan city and the rooms of the archaeological museum, interacting and dialoguing with other characters in the game. It is a side-scrolling 2D narrative game, which explores feelings such as love, dreams, fear, where the player takes the role of a son who intends to find his archaeologist dad. One of the main features of this game is its hand-painted graphics made by the British artist Sean Wenham.

During the first lockdown in Spain, the Department of Tourism of the Madrid City Council launched a cultural game on its official pages. Players from all over the world had the opportunity to test themselves, to see how well they knew the secrets of the Spanish capital.

In Denmark, the Faroe Islands implemented the "Remote tourism" project between April 2020 and July 2020. While it was not a game in the real sense, it required users to behave like gamers. In fact, they had to use a real virtual joypad to cross the islands thanks to the video recorded by the islanders with Go-pro mounted on foot, by boat or by helicopter. Using the controls on the joypad, users could also choose how to move during the tour, whether to turn, walk, run and even jump.

In Asia, Chinese cultural institutions released a series of educational games and interactive digital artwork online during the 2020 Lunar New Year. One of the most active centers was the Suzhou Museum which is the main home to traditional Chinese painting. It has launched four mini games on its WeChat and Weibo official accounts through which users have had the opportunity to get in touch and interact with different art works. For example, the online game "DIY Landscape Painting" allowed users to create digital paintings by using some of the most striking classical elements of Chinese painting, such as endless rivers, thatched cottages and herons.

5.5 The main solutions for holding large-scale events

Events are deeply rooted in human culture. They are moments in which society manifests itself and makes its own culture and identities visible in a spectacularized forms (Zeppel and Hall, 1991; Weiler and Hall, 1992). Even from a tourist point of view, over time, events have assumed an increasingly central role in terms of tourist attractions. In fact, they have become an important dynamic product of the tourism and entertainment industry capable of attracting a large number of visitors around them. The success of events derives both from the temporariness and uniqueness of each event, which take place in a specific destination at a given moment, and from the atmosphere that characterizes each initiative (Getz, 1997).

The range of events is very wide. It includes a large number of elements, from sporting events of great international appeal to more intimate ceremonies

(Shone and Parry, 2004). Some initiatives have now become fixed appointments. They are repeated over time, offering both elements of continuity with the past and elements of innovation to amaze the participants and offer something new and different from time to time, but still in line with previous sessions.

More specifically, an event has a tourist characterization since it not only meets specific demands (e.g., sporting, cultural or religious), but it increases the degree of attractiveness and the added value of the host location as a tourist destination (e.g., Mossberg, 2000; Dwyer, Forsyth and Spurr, 2005).

This analytical interpretation allows us to argue that events during a tourist vacation can be defined as initiatives that encourage physical mobility for entertainment and escape from the daily routine, but at the same time offer the opportunity to connect with the social and cultural heritages of the individual territories, bringing together travelers and local residents. Thanks to events, people have the opportunity to meet each other and deepen the reciprocal understanding through direct contact and exchange.

When the pandemic manifested itself in all its violence, the main events of international significance were postponed with the hope that rescheduling was only a temporary plan. This response was taken in all those cases in which the organizers refused to find alternative solutions or misunderstood the nature of the event.

Exactly for this reason, some initiatives were suspended. For example, the Coachella Valley Music and Arts Festival, one of the most important musical events in the United States, took place every year around the end of April in the Empire Polo Fields of Indio in California with hundreds of thousands of people from various countries. Thanks to this event, the Californian desert has become an iconic place with great tourist popularity. The 2020 Festival, however, was postponed from April to October and then cancelled altogether, along with its sister festival, Stagecoach, by Riverside public health officials to protect the health of the local com, because the COVID-19 contagion curve did not go down. The same decision was made for 2021 as well.

Similarly, in China, in the pre-COVID era, the Chinese New Year, which coincides with the beginning of the Spring Festival, was the busiest time of travel of a year. Each year it caused mass travel throughout China. As we mentioned in more detail in previous chapters, the Chinese government canceled several large Chinese New Year celebration events in 2020. In that same period the authorities imposed strict lockdowns and ordered the suspension of flights and connections among cities by public transport a few hours before the start of the major events, on January 25. In addition, to compensate for the restrictions, local authorities in many cities promised incentive coupons to the local residents. For example, in Beijing the authorities encouraged companies to pay overtime and offer free meals to the employees who chose not to return to their home cities, while in Hangzhou, not far from Shanghai, workers who did not travel were promised a bonus worth around 130 euros. In other cities, local authorities rewarded people who had not left with gift baskets or easier access to public and health services.

In addition, some of the most important events were held with no audience or on live-streaming platforms with audiences cheering only on screen.

Among the most special images of this type, in terms of its exceptionality and symbolism, we cannot forget Pope Francis who, crossed an empty St. Peter's Square in Rome alone in the rain, toward Eucharistic Adoration on March 27, 2020, the day of the *Urbi et orbi*.

The images of this blessing went viral on social media around the world, because it bore not only significance in the context of the Catholic religion, but had a strong media impact on the world population for the iconographic and narrative synthesis of the consequences of the pandemic.

The emptiness of the square emphasized the impossibility of physical aggregation among believers. Spectators from all over the world interested in the event had the opportunity to watch it through live broadcasting on television or on the web, in a private space that guaranteed the enjoyment of the event beyond the feeling of the public space. The monumental void of St. Peter's Square symbolically represents the forced redefinition of participation in events.

Likewise, other events have also been adapted to meet the needs of the moment, thus being given a new definition.

For example, the 65th Eurovision Song Contest was held at Ahoy Rotterdam in the Netherlands, on May 12, 14 and 16, 2020. On March 18, 2020, the European Broadcasting Union announced the cancellation of the event due to the pandemic, rescheduling it for 2021. Following this unprecedented cancellation, the organization worked to offer an alternative remote event entitled "Eurovision: Europe Shine a Light." Each artist performed the song that they selected to join the competition, so as to offer an alternative show to spectators from all over the world.

A similar situation occurred in the fashion industry. In June, 2020, for London fashion week, a multimedia platform was launched by the British Fashion Council. This included interviews, podcasts and digital showrooms open to the public. The digital transformation of the London Fashion Week was also an opportunity to create gender-neutral fashion shows for the first time, combining men's and women's fashion collections.

Connection and creativity, together with inclusion thus became the keywords for change.

Similarly, in July 2020, the first "digital" Milan Fashion Week debuted to present the men's collections and the men's and women's precollections for Spring–Summer 2021.

In 2021, the live streaming of runway was further enriched. The platform created by the National Chamber of Italian Fashion not only streamed the fashion shows of the brands, but it also offered some value-added functions, such as videos, streaming events and thematic rooms to explore topics such as sustainability and inclusiveness in fashion. The inaugural party was broadcast live on Instagram. In addition, the fashion week also appeared on the video social community TikTok, which was chosen to host many online thematic initiatives and offer insights related to the main event.

Other traditional events have found a digital solution too, such as the Christmas markets around the world. Many cities presented a new tradition, offering online what they had previously offered up close. Already in the second half of 2020, many colorful websites were set up on which sellers and artisans exhibited their products and gave people the opportunity to order and purchase them from home. Certainly this experience appears in some ways limited. The typical smells and flavors of Christmas markets did not have the opportunity to go outside the devices. However, this was a solution to allow people to experience the atmosphere of the local Christmas markets in an alternative way.

Finally, the spread of COVID-19 has also spurred the creation of new events, which have been able to entertain people at home due to the pandemic. Among these, the most famous were the charity shows organized by Global Citizen, with the patronage of the European Commission in support of the World Health Organization, which were created with the aim of promoting the practice of physical distancing during the pandemic and to raise funds in support of the most marginalized underdeveloped communities in the battle against COVID-19.

The first event was called "One World: Together at Home." It was broadcast live on April 18, 2020 on major social media platforms such as YouTube and Facebook and on major national television channels globally. It was a music and entertainment show that lasted over 6 hours for the fund-raising of the Global Goal: Unite for Our Future campaign. The project managed to raise $127.9 million globally.

Two months later the "Global Goal: Unite for Our Future—The Concert" event was aired. Broadcasted on June 27, 2020 in over 180 countries, it collected overall donations of $6.9 billion from companies, foundations and governments, including all those of the G7, to finance research against COVID.

The event managed to combine a musical show with the performance in some beautiful places in the world. Jennifer Hudson, the first artist to perform, sang "Where Peaceful Waters Flow" aboard a boat on the Chicago River, while Shakira and her band performed the song "Sale el sol," on a rooftop showing a charming view of the city of Barcelona in Spain. Another celebrity involved in Global Goal was Miley Cyrus with her cover of "Help!," performed in the middle of an empty stadium, the Rose Bowl in Pasadena, California, using Dolly Parton's version of the Beatles' 1965 classic. To make the message even clearer, the word "Help!" was written in giant letters on the playing field, with the performer singing inside the exclamation point.

The first event that marked the progressive recovery to normalcy and mobility was the 2021 Super Bowl, which is the final championship of the

American National Football League. Raymond James Stadium in Tampa, Florida welcomed a small number of spectators. Of the normal-capacity 65,000 seats, only 25,000 were occupied by spectators, and 7,500 by vaccinated doctors and nurses from all over the country. They were invited as guests of honor, and the sponsor also wanted to show gratitude for their commitment in the fight against the pandemic. Most of the seats in the stadium were filled with cardboard silhouettes, with images of celebrities such as Eminem, Drake, Billie Eilish, together with photos of ordinary spectators who paid $100 for the honor of being present in the stands virtually, with their own photo printed on a cardboard silhouette.

This was a new practice that will probably be replicated in the future too. Some people not only watched the event from home, but a photographic reproduction of them was present on site. This enabled stay-at-home people to still be able to say "I was present at that event" even if in a transfigured way.

A similar initiative has been carried out also in China. During the national volleyball championship matches, fans could sign up to be present on screen of the stadium, yelling and cheering on the team they support. Their faces showed up on the LED screen in the stadium, where no audience was present due to national health regulations. These audiences, at least 20 per match, who were "present at the stadium" but actually stayed at home, are called "cloud audiences." They could connect with the stadium by a special system of 5G real-time communication designed by "Miguvideo.com." Even retired volleyball stars could interact with these audiences.

From a social point of view, this new type of event managements makes it possible to detect further forms and interpretations of the concept of presence and participation in events.

Conclusions

As virtuality began to come into our lives more and more, it was criticized for being considered a kind of static and meaningless simulation of reality. However, with the advancement of technology it has been shown to provide sensory and emotional experiences through simulation (Melotti, 2011). Consequently, virtual environments have become, especially for the new generations, a new trend of instantaneity. This visual culture is able to provide unique experiences, sometimes anticipating the fruition of reality.

On many other occasions, however, it replaces the real world, particularly when the latter is not accessible.

Thanks to the presence of an increasing number of technological developers, during the pandemic we have witnessed a proliferation of virtual mobility solutions, which has temporarily met, at least in part, many people's desire to travel. I would like to briefly review the advantages and possibilities that this type of practice has in general and that, in particular, it has brought to the people during the period of the pandemic.

First of all, stationary tourism has the merit of allowing people all over the world to anticipate their tourist experience. In other words, travelers take a mediated approach to the tourist experience of completing a trip by implementing the activities they would have undertaken in the real world. In this sense, stationary tourism can stimulate their intention and help them with their decision-making, offering a simulated preview of the experience (e.g., Cho, Wang and Fesenmaier, 2002; Huang *et al.*, 2016; Kim, Lee and Preis, 2020; Rahimizhian, Ozturen and Ilkan, 2020; El-Said and Aziz, 2021).

Secondly, another connection between tourism and virtuality concerns the sphere of marketing. People who practice one of the many forms of stationary tourism can enjoy in advance the goods, products and services that may be of interest to them, through a simulated exploration of a territory. This experience can make them more aware as consumers, because they are able to identify in advance what items to buy, on the spot or before actual departure, to meet their needs: food to eat, facilities to stay in, attractions to visit, means of transport to take, tickets to buy, and so on (e.g., Jung and Dieck, 2018; Leung, Lyu and Bai, 2020; Lin, Huang and Ho, 2020; Zeng *et al.*, 2020). From this analysis, it is safe to argue that stationary tourism acts as a forerunner of moving tourism, and it can also be interpreted as a means of freeing the tourist from false expectations and the possibility of holiday disappointment.

Furthermore, stationary tourism is capable of satisfying different needs. It can perform both entertainment functions and have cognitive purposes. In fact, new technologies make it possible to effectively collect an infinite series of information through a type of interactive, sensory and participatory learning.

Similarly, some scholars (e.g., Hobson and Williams, 1995; Ford, 2001; Goodall *et al.*, 2004; Kim and Hall, 2019) have emphasized that stationary tourism can also be a tool for ensuring an increase in the accessibility of tourist sites. In fact, practicing online tourism allows people to visit places that are too distant, too inhospitable, or too fragile, or to touch articles of enormous historical value that are often prohibited from being physically touched. As I have written elsewhere (Monaco, 2019), today there are several tourism barriers which can be tangible, but also sociocultural in nature. Many territories, for example, are hostile toward some sexual and gender minorities (such as transgender or homosexual people). Or other places in the world have architectural barriers which are not so friendly to the people with physical disabilities. There are also economic barriers, however, for example, tourist attractions which are open exclusively to the rich. In this respect, stationary tourism is certainly more inclusive and universal, because it makes the tourist experience equitable, irrespective of their social or economic class.

In this sense, stationary tourism allows people to be able to enjoy a 2.0 form of the "right to the city," in Lefebvrian terms, since it places all stationary tourists in the same conditions. With this expression, Lefebvre (1968) referred both to access to the resources of the city and to the possibility of maintaining significant social and economic relationships.

Furthermore, from this perspective, stationary tourism also favors a more sustainable tourism, reducing the ecological impact of visitors (e.g., Dewailly, 1999; Kask, 2018; Kwok and Koh, 2021). It can therefore be considered a kind act not only in terms of cultural and territorial promotion, but also for its potential as a proctor of nature. Furthermore, it helps open historical and naturalistic sites to the public, while strengthening historical preservation. At the same time, sustainable tourism, in its true sense, does not only look to nature, but also to the well-being of local inhabitants. Thus, making more and more sites virtually available can also be a strategy to decongest the overcrowding in some tourist attractions, freeing them from the overtourism that in the past has affected various destinations, such as cities of art and seaside resorts.

Another positivity of these experiences is the capability of solving a series of problems that a traveler may face during a moving tourism. These problems include, for example, waiting, stationary in long lines to enter into the buildings, the documents to be shown, the possible language barriers, the exchange of currency and many surprises.

In this particular historical period, while awaiting the full recovery of international mobility in safety, destinations around the world have grasped the opportunity of virtual tours to promote their artistic, cultural and urban heritage, involving a very large area of attraction (e.g., Atsiz, 2021). Likewise, since moving tourism is suffering from severe limitations, stationary tourism has made to satisfy other needs in addition to those that have just been listed.

In fact, new needs and problems have emerged in accordance with the critical issues of the pandemic period. Thus, stationary tourism has made it possible to overcome the ban on physical mobility imposed by individual governments. Sociologically, it is safe to argue that people had the opportunity to pop their heads out of their homes even while they were in lockdown, practicing different forms of travel that have allowed them to escape from the confided spaces and thus (virtually) enjoy the world's tourist heritage.

More specifically, in the social context of the pandemic, nonphysical travel has also served as a remedy to combat the boredom during lockdowns, when people were unable to leave the house and thus had to find alternative ways to pass the long time (e.g., Monaco, 2020; Yang *et al.*, 2021). Some research on the subject (e.g., Kerawalla *et al.*, 2006; Stangl and Weismayer, 2008; Li, Song and Guo, 2021) has highlighted that this type of experience allowed for a temporary escape from reality, so that it could be considered a possible outlet for real-life frustrations.

In addition, stationary tourism has made it possible to travel safely, satisfying the need for tourism without exposing tourists to the risk of contagion. Certainly the initiatives that have been carried out have allowed millions of citizens globally to visit many tourist places from the safety of their own homes, by themselves or in company, crossing the boundaries which were closed because of the virus, even if virtually.

Thus, physical interactions have been avoided, without forfeiting contact with the destination's local population. In other words, this type of action has made it possible to create contact with local communities, reminding them that one day they will be able to see in person what they have already experienced remotely through texts, videos, photos and audio files.

As we have seen in the previous chapters, tourists find it possible on different online platforms to communicate with other people and exchange messages and information with them, for example, video tours on some portals such as YouTube, and online gaming. In addition, during the dematerialized tourist experience, people can communicate through instant messaging, to exchange opinions, feelings, ideas, advice and to comment on what they see on many social networking platforms.

It is not possible to predict where the development of virtual reality will lead us, nor if it will actually be considered one day as a completely alternative form to moving tourism. Like any other form of emerging solution, stationary tourism presents both problems and opportunities. Only through research and experience will it be possible to verify more concretely if this could change into a resource.

After all, even when the vaccination campaign is completed, people may be reluctant to travel and it may be some time before travel restrictions are removed. This means that stationary tourism experiences will retain their favorable position at least until for a few year. The more people feel comfortable with this type of tourism experience, the more they will investigate other destinations in the world.

Therefore, the hypothesis that stationary tourism could represent the new norm of cultural enjoyment is not too far-fetched.

A further possible scenario is that gradual hybrid forms of digital and real experiences can characterize tourist trips. In other words, a "phygital" experience (e.g., Neuburger, Beck and Egger, 2018; Ballina, Valdes and Del Valle, 2019; Agostino, Arnaboldi and Lorenzini, 2020) in which stationary tourism and moving tourism alternate, occur and overlap each other could represent one of the many new postpandemic revolutions.

From a sociological point of view, the so-called "never-ending tourism" appears to be a new trend in the near future. It has been conceptualized as an approach to the tourist experience with no limitations on visiting places. Quite the opposite: it begins before the physical journey and it continues after it, thanks to the use of new communication technologies, which allow people to create lasting bonds with destinations. Never-ending tourism actually implies an expansion in space and time of the tourist experience. According to data produced by the Digital Innovation in Tourism Observatory of the School of Management of the Politecnico of Milan, 40 percent of Internet users increased the time dedicated to video entertainment in 2020. Consequently, many territories have taken steps to derive value from this novelty, opening new channels of communication with tourists and offering products and services even at a distance, waiting for meeting visitors face to face. At the same time, the expansion of the tourist experience guaranteed by never-ending tourism evidently extends the demand for tourism well beyond the traditional holiday season.

Bibliography

Abulibdeh, A (2020) 'Can COVID-19 mitigation measures promote telework practices?, *Journal of Labor and Society*, 23, 4: 551–576.

Agostino, D, Arnaboldi, M and Lorenzini, E (2020) 'Verso un new normal dei musei post-COVID 19: quale ruolo per il digitale?', *Economia della Cultura*, 1: 79–83.

Alčaković, S, Pavlović, D and Popesku, J (2018) 'Millennials and gamification. A model proposal for gamification application in tourism destination', *Marketing*, 48: 207–214.

Andreoli, M (2018) 'La Realtà Virtuale al servizio della Didattica', *Studi sulla Formazione*, 21: 33–56.

Atsiz, O (2021) 'Virtual reality technology and physical distancing: A review on limiting human interaction in tourism', *Journal of Multidisciplinary Academic Tourism*, 1: 27–35.

Azuna, R T (1997) 'A survey of augmented reality', *Presence: Teleoperators and Virtual Environments*, 6, 4: 355–385.

Baer, M W and Hui, J K (2019) 'How interactive video games helped shape the modern computer world', in AA.VV *2019 6th IEEE History of Electrotechnology Conference (HISTELCON)*, Glasgow: IEEE.

Ballina, F J, Valdes, L and Del Valle, E (2019) 'The phygital experience in the smart tourism destination', *International Journal of Tourism Cities*, 5, 4: 656–671.

Baudrillard, J (1997) *Art & Artefact*, Leicester: Sage.

Benjamin, W (1936) *Das Kunstwerk im Zeitalter seiner technischen Reproduzierbarkeit*, Berlin: Suhrkamp Verlag.

Berné, C, García-González, M, García-Ucedac, M E and Múgica, J M (2015) 'The effect of ICT on relationship enhancement and performance in tourism channels', *Tourism Management*, 48: 188–198.

Bonesio, L (2002) *Oltre il paesaggio*, Casalecchio: Arianna editrice.

Brown, B (2017) 'The Psychology of Gamification in 2016: Why It Works (& How To Do It!)', Retrieved from *Bitcatcha*: https://www.bitcatcha.com/blog/2016/gamify-website-increase-engagement/ (accessed 04/02/2021).

Brunelli, A (2017) 'Vedere l'invisibile: musei e biblioteche nell'era della Realtà Aumentata', *Bibliotime*, 1, 2.

Buhalis, D and Licata, M C (2002) 'The future eTourism intermediaries', *Tourism Management*, 23, 3: 207–220.

Bulencer, P and Egger, R (2015) *Gamification in Tourism, Designing Memorable Experiences*, Norderstedt: Bod: Books on Demand.

Burdea, G C and Coiffet, P (2003) *Virtual Reality Technology*, New York: John Wiley & Sons.

Carmigniani, J, Furht, B and Anisetti, M *et al.* (2011) 'Augmented reality technologies, systems and applications', *Multimedia Tools and Applications*, 51: 341–377.

Carpenter, D and Dunn, J (2020) 'We're all teachers now: Remote learning during COVID-19', *Journal of School Choice*, 14, 4: 567–594.

Castells, M (1996) *The Rise of the Network Society*, Malden: Blackwell Publishers.

Cheong, R (1995) 'The virtual threat to travel and tourism', *Tourism Management*, 16, 6: 417–422.

Cho, Y H, Wang, Y and Fesenmaier, D R (2002) 'Searching for experiences: The web-based virtual tour in tourism marketing', *Journal of Travel & Tourism Marketing*, 12, 4: 1–17.

Consalvo, M (2006) 'Console video games and global corporations: Creating a hybrid culture', *New Media & Society*, 8, 1: 117–137.

Cox, C, Burgess, S, Sellito, C and Buultjens, J (2009) 'The role of user generated content in tourists' travel planning behaviour', *Journal of Hospitality Marketing and Management*, 18, 8: 743–764.

Cruz-Neira, C, Sandin, T A and De Fanti, R V (1993) 'Surround screen projection-based virtual reality: The design and implementation of the cave', in M C Whitton (ed.) SIGGRAPH '93: *Proceedings of the 20th Annual Conference on Computer Graphics and Interactive Techniques*, New York: Association for Computing Machinery.

De Paolis, L M (2012) 'Applicazione interattiva di realtà aumentata per i beni culturali', *Scientific Research and Information Technology*, 2, 1: 121–132.

Dewailly, J M (1999) 'Sustainable tourist space: From reality to virtual reality?', *Tourism Geographies*, 1: 41–55.

Dwyer, L, Forsyth, P and Spurr, R (2005) 'Estimating the impact of special events on an economy', *Journal of Travel Research*, 43, 4: 351–359.

Earnshaw, R A, Gigante, M A and Jones, H (1993) *Virtual Reality Systems*, San Diego: Academia Press.

El-Said, O and Aziz, H (2021) 'Virtual tours a means to an end: An analysis of virtual tours' role in tourism recovery post COVID-19', *Journal of Travel Research*, 45.

Ford, P J (2001) 'Paralysis lost: Impacts of virtual worlds on those with paralysis', *Social Theory and Practice*, 27, 4: 661–680.

Fulco, I (2006) *Virtual geographic. Viaggi nei mondi dei videogiochi*, Milan: Costa & Nolan.

Furht, B (2011) *Handbook of Augmented Reality*, Boca Raton: Springer.

Getz, D (1997) *Event Management & Event Tourism*, New York: Cognizant Communication Corporation.

Goodall, B, Pottinger, G, Dixon, T and Russell, H (2004) 'Heritage property, tourism and the UK disability discrimination act', *Property Management*, 22, 5: 345–357.

Granieri, G (2006) *La società digitale*, Bari: Laterza.

Gros, F (2009) *Marcher, une philosophie*, Paris: Carnet Nord.

Gruescu, R, Nanu, R and Pirvu, G (2009) 'Information and communications technology and internet adoption tourism', *Bulletin UASVM Horticulture*, 66, 2: 407–413.

Gutierrez, M, Vexo, F and Thalmann, D (2008) *Stepping into Virtual Reality*, London: Springer.

Guttentag, D A (2010) 'Virtual reality: Applications and implications for tourism?', *Tourism Management*, 31, 5: 637–651.

Heeter, C (1992) 'Being there: The subjective experience of presence', *Presence*, 1, 2: 262–271.

Hobson, J S P and Williams, A P (1995) 'Virtual reality: A new horizon for the tourism industry', *Journal of Vacation Marketing*, 1, 2: 125–136.

Huang, Y C, Backman, K F, Backman, S J and Chang, L L (2016) 'Exploring the implications of virtual reality technology in tourism marketing: An integrated research framework', *International Journal of Tourism Research*, 18: 116–128.

Hudson, S and Ritchie, J R B (2006) 'Promoting destination via film tourism: An empirical identification of supporting marketing initiatives', *Journal of Travel Research*, 44: 387–396.

Hwang, S and Yu, D (2012) 'GPS localization improvement of smartphones using built-in sensors', *International Journal of Smart Home*, 6, 3: 1–8.

Jacobson, J and Holden, L (2005) *The Virtual Egyptian Temple. World Conference on Educational Multimedia Hypermedia & Telecommunications*, Montreal: ED MEDIA.

Jaeger, L, Zarb, N and David, A (2020) *Global Gaming Study: More Gamers Spending More Money in COVID Lockdowns. Which Publishers Will Benefit?*, Bonn: Simon Kucher & Partners.

Jalla, D (2019) 'Comunicare e conservare al tempo del web', in S D Orlandi, G Calandra, V Ferrara, A M Marras and S Radice (eds.) *Web Strategy Museale. Monitorare e progettare la comunicazione culturale nel web*, Milan: ICOM.

Jansson, A (2002) 'Spatial phantasmagoria the mediatization of tourism experience', *European Journal of Communication*, 17, 4: 429–443.

Jenkins, F and Smith, J (2021) 'Work-from-home during COVID-19: Accounting for the care economy to build back better', *The Economic and Labour Relations Review*, 32, 1: 22–38.

Jones, S (2006) 'Reality and virtual reality', *Cultural Studies*, 20, 2: 211–226.

Jung, T and Dieck, M C T (eds.) (2018) *Augmented Reality and Virtual Reality: Empowering Human, Place and Business*, Berlin: Springer-Verlag.

Kask, S (2018) *Virtual Reality in Support of Sustainable Tourism. Experiences from Eastern Europe*, Tratu, Estonia: Estonian University of Life Sciences Press.

Kenway, J and Epstein, D (2021) 'The covid-19 conjuncture: Rearticulating the school/home/work nexus', *International Studies in Sociology of Education*, 1.

Kerawalla, L, Luckin, R, Seljeflot, S and Woolard, A (2006) 'Making it real: Exploring the potential of augmented reality for teaching primary school science', *Virtual Reality*, 10: 163–174.

Kim, M J and Hall, C M (2019) 'A hedonic motivation model in virtual reality tourism: Comparing visitors and non-visitors', *International Journal of Information Management*, 46: 236–249.

Kim, M J, Lee, C K and Preis, M W (2020) 'The impact of innovation and gratification on authentic experience, subjective well-being, and behavioral intention in tourism virtual reality: The moderating role of technology readiness', *Telematics and Informatics*, 49: 16.

Koltko-Rivera, M E (2005) 'The potential societal impact of virtual reality', in *ERCIM Working Group Proceedings of HCI International 2005, The 11th International Conference on Human Computer Interaction, Volume 9—Advances in Virtual Environments Technology: Musings on Design, Evaluation, and Applications*, Las Vegas: Mira Digital Publishing.

Kwok, A O and Koh, S G M (2021) 'COVID-19 and extended reality (XR)', *Current Issues in Tourism*, 24, 14: 1935–1940.

Lee, S, Tewolde, G and Kwon, J (2014) 'Design and implementation of vehicle tracking system using GPS/GSM/GPRS technology and smartphone application', in R Minerva (ed.) *2014 IEEE World Forum on Internet of Things (WF-IoT)*, Seoul: IEEE.

Lefebvre, H (1968) *Le Droit à La Ville*, Paris: Anthropos.

Leung, X Y, Lyu, J Y and Bai, B A (2020) 'Fad or the future? Examining the effectiveness of virtual reality advertising in the hotel industry', *International Journal of Hospitality Management*, 88: 102391.

Lew, A A, Hall, C M and Williams, A M (2014) *The Wiley Blackwell Companion to Tourism*, Hoboken: John Wiley & Sons.

Li, Y, Song, H and Guo, R (2021) 'A study on the causal process of virtual reality tourism and its attributes in terms of their effects on subjective well-being during COVID-19', *International Journal of Environmental Research and Public Health*, 18, 3: 1019.

Lin, L P, Huang, S C and Ho, Y C (2020) 'Could virtual reality effectively market slow travel in a heritage destination?', *Tourism Management*, 78: 104027.

Macionis, N (2004) 'Understanding the film-induced tourist', in W Frost, W C Croy and S Beeton (eds.) *Proceedings of the International Tourism and Media Conference*, Melbourne: Monash University.

Main, H C (2002) 'The expansion of technology in small and medium hospitality enterprises with a focus on net technology', *Information Technology & Tourism*, 4: 167–174.

Mandelli, A and Accosto, C (2012) *Social mobile marketing: l'innovazione dell'ubiquitous marketing con device mobili, social media e realtà aumentata*, Milan: Egea.

Mandich, G (1996) *Spazio-tempo. Prospettive sociologiche*, Milano: FrancoAngeli.

Manovich, L (2001) *The Language of New Media*, Cambridge: MIT Press.

Manovich, L (2013) *Software Takes Command*, New York: Bloomsbury Academic.

Matthewman, S and Huppatz, K (2020) 'A sociology of covid-19', *Journal of Sociology*, 56, 4: 675–683.

Melotti, M (2011) *The Plastic Venuses. Archaeological Tourism and Post-Modern Society*, Newcastle: Cambridge Scholars.

Meschini, A (2011) 'Tecnologie digitali e comunicazione dei beni culturali. Stato dell'arte e prospettive di sviluppo', *Disegnare Con*, 4: 14–24.

Meyrowitz, J (1985) *No Sense of Place: The Impact of the Electronic Media on Social Behavior*, Oxford: Oxford University Press.

Meyrowitz, J (2005) 'The rise of glocality: New senses of place and identity in the global village', in K Nyìri (ed.) *The Global and the Local in Mobile Communication*, Wien: Passagen Verlag.

Milgram, P and Kishino, F (1994) 'A taxonomy of mixed reality visual displays', *IEICE Transactions on Information System*, 77, 12: 1321–1329.

Mohanty, P, Hassan, A and Ekis, E (2020) 'Augmented reality for relaunching tourism post-COVID-19: Socially distant, virtually connected', *Worldwide Hospitality and Tourism Themes*, 1, 6: 753–760.

Molz, G J (2004) 'Playing online and between the lines: Round-the-world websites as virtual places to play', in M Sheller and J Urry (eds.) *Tourism Mobilities: Places to Play, Places in Play*, London: Routledge.

Monaco, S (2019) *Sociologia del Turismo Accessibile. Il Diritto alla Mobilità e alla Libertà di Viaggio*, Varrazze: PM Editore.

Monaco, S (2020) 'Turismo in lockdown. Tra misure economiche e politiche simboliche', *Rivista Trimestrale di Scienze dell'Amministrazione. Studi di Teoria e Ricerca Sociale*, 2: 1–18.

Monaco, S (2015) 'Società 2.0: trasformazioni comportamentali, artistiche e linguistiche nell'era di Internet', *FUTURI*, 5: 47–57.

Montagna, L (2018) *Realtà virtuale e realtà aumentata. nuovi media per nuovi scenari di business*, Milan: Hoepli.

Montola, M, Stenros, J and Waern, A (2009) *Pervasive Games. Theory and Design*, Burlington: Morgan Kaufmann.

Mossberg, L L (ed.) (2000) *Evaluation of Events Scandinavian Experiences*, New York: Cognizant Communication Corporation.

Neuburger, L, Beck, J and Egger, R (2018) 'The phygital tourist experience: The use of augmented and virtual reality in destination marketing', in M A Camilleri (ed.) *Tourism Planning and Destination Marketing*, Bingley: Emerald Publishing Limited.

Nuvolati, G (2013) *L'interpretazione dei luoghi. Flânerie come esperienza di vita*, Florence: Florence University Press.

Olszewski, R, Pałka, P and Turek, A (2018) 'Solving smart city transport problems by designing carpooling gamification schemes with multi-agent systems: The case of the so-called Mordor of Warsaw', *Sensors*, 18, 1: 141.

Pecchinenda, G (2010) *Videogiochi e cultura della simulazione. La nascita dell'homo game*, Bari: Laterza.

Rahimizhian, S, Ozturen, A and Ilkan, M (2020) 'Emerging realm of 360-degree technology to promote tourism destination', *Technology in Society*, 63: 101411.

Refsland, S T, Ojika, T, Addison, A C and Stone, R (2000) 'Virtual heritage: breathing new life into our ancient past', *IEEE Multimedia*, 7, 2: 20–21.

Rheingold, H (1991) *Electropolis: Communication and Community on Internet Relay Chat*, Melbourne: University of Melbourne.

Riley, R and Van Doren, C S (1992) 'Movies as tourism promotion: A 'Pull' factor in a 'Push Location', *Tourism Management*, 13, 3: 267–274.

Seaborn, K (2015) 'Gamification in theory and action: A survey', *International Journal of Human-Computer Studies*, 1: 7414–7431.

Shone, A and Parry, B (2004) *Successful Event Management: A Practical Handbook*, Boston: Thomson Learning.

Skinner, H, Sarpong, D and White, G (2018) 'Meeting the needs of the millennials and generation z: Gamification in tourism through geocaching', *Journal of Tourism Futures*, 4, 4.

Soteriades, M, Aivalis, C and Varvaressos, S (2004) 'e-Marketing and e-commerce in the tourism industry: A framework to develop and implement business initiatives', *Tourism Today*, 4: 1–18.

Stangl, B and Weismayer, C (2008) 'Websites and virtual realities: A useful marketing tool combination? An exploratory investigation', in P O'Connor, W Höpken and U Gretzel (eds.) *Information and Communication Technologies in Tourism*, New York: Springer.

Steinicke, F (2016) *Being Really Virtual. Immersive Natives and the Future of Virtual Reality*, Cham: Springer.

StreamElements (2021) *State of the Stream*, Tel Aviv: StreamElements.

Sudharshan, D (2020) *Augmented Reality. Marketing in Customer Technology Environments*, Bingley: Emerald Publishing.

Tanuvi, J (2020) 'Virtual Reality Travel Is Changing The Tourism Industry, But Is It Being Responsible?', *The Dope*, 11/20/2020.

Terlutter, R and Capella, M L (2013) 'The gamification of advertising: Analysis and research directions of in-game advertising, advergames, and advertising in social network games', *Journal of Advertising*, 2, 3: 95–112.

Thomas, S and Mintz, A (eds.) (1998) *The Virtual and the Real. Media in the Museum*, Washington D.C.: American Association of Museums.

Trinidad, J E (2021) 'Equity, engagement, and health: School organisational issues and priorities during COVID-19', *Journal of Educational Administration and History*, 53, 1: 67–80.

Umaschi Bers, M (2010) 'Let the games begin: Civic playing on high-tech consoles', *Review of General Psychology*, 14, 2: 147–153.

UNWTO (2020) *UNWTO Report: 10th Report on Travel Restrictions*, Madrid: UNWTO.

Urry, J (2002) 'Mobility and proximity', *Sociology*, 36, 2: 255–274.

van Dijk, J (1999) *The Network Society: Social Aspects of New Media*, London: Sage.

Wagler, A and Hanus, M D (2018) 'Comparing virtual reality tourism to real-life experience: Effects of presence and engagement on attitude and enjoyment', *Communication Research Reports*, 35, 5: 456–464.

Weber, J (2013) 'Gaming and Gamification in Tourism'. Retrieved from *Think thank*: https://thinkdigital.travel/wp-content/uploads/2014/05/Gamification-in-Tourism-Best-Practice.pdf (accessed 04/02/2021).

Weiler, B and Hall, C M (eds.) (1992) *Special Interest Tourism*, London: Belhaven.

Williams, P and Hobson, J S P (1995) 'Virtual reality and tourism: Fact or fantasy?', *Tourism Management*, 16: 423–427.

Woods, S (2012) *Eurogames: The Design, Culture and Play of Modern European Board Games*, Jefferson: McFarland & Company.

Xu, F, Buhalis, D and Weber, J (2017) 'Serious games and the gamification of tourism', *Tourism Management*, 60: 244–256.

Yang, T, Lai, I, Fan, Z B and Mo, Q M (2021) 'The impact of a 360° virtual tour on the reduction of psychological stress caused by COVID-19', *Technology in Society*, 64: 101514.

Yates, A, Starkey, L, Egerton, B and Flueggen, F (2020) 'High school students' experience of online learning during Covid-19: The influence of technology and pedagogy', Technology, Pedagogy and Education, 30, 1: 59–73.

Zeng, G J, Cao, X N, Lin, Z B and Xiao, S H (2020) 'When online reviews meet virtual reality: Effects on consumer hotel booking', *Annals of Tourism Research*, 81: 102860.

Zeppel, H and Hall, C M (1991) 'Selling art and history: Cultural heritage and tourism', *The Journal of Tourism Studies*, 2, 1: 29–45.

Zhou, Z (2004) *E-Commerce and Information Technology in Hospitality and Tourism*, New York: Delmar Publishing.

Žižek, S (2020) *Pandemic! Covid-19 Shakes the World*, New York: OR Books.

Conclusion

A look to the future

Unlike previous tourism crises, which limited travel at national or regional levels, the COVID-19 pandemic not only represented a setback for the industry around the world, but implicitly introduced new visions, new needs and new habits to the field of tourism. The uniqueness of this pandemic is that it has plagued the entire world, presenting a completely new situation: first of all it has affected every person in every place in the world, and then it also directly has affected a lot of sectors and industries, including the entire tourist chain.

Furthermore, the pandemic has exerted a negative impact on the economy which was already very fragile, thus further aggravating the uncertain scenario of continuing geopolitical, social and commercial tensions on a global level. Under these circumstances, the spread of COVID-19 can be defined as one of the greatest threats and challenges that the world of tourism has ever faced.

As Zwanka and Buff (2021) explained, COVID-19 is only the latest of many revolutionary or traumatic events that have been able to influence the definition of generational identities powerfully, just as the great 1960s economic boom, world wars, but also the assassination of President Kennedy or the birth of the Internet had done in the past. In other words, being a catastrophe, the pandemic is typical of an event which will accelerate social change.

Not surprisingly, another generation called "coronials" (Harmony, 2020) or "quaranteens" (ET Bureau, 2020) has also been added to the traditional list of generations. This term refers not only to the people who were born during the period of the pandemic, but also, in a broader sense, to the adolescents of this particular historical period, whose values and identities as people and tourists have been strongly forged by this catastrophe. According to Attias-Donfut (1991) the concept of generation has four dimensions: demographical; genealogical and familial; historical; and sociological, similar to the notion of a "generational cohort." This notion is one of the most important sociological concepts used to explain social change (Ward, 1974; Holbrook and Schindler, 1989; Alwin and McCammon, 2007; Ladwein, Carton and Sevin, 2009) and to measure each historical era. Furthermore, since the new millennium, there has been a growing academic

DOI: 10.4324/9781003195177-7

interest in generational analysis in tourism studies. The most recent studies have shown that tourist behaviors are also different on a generational basis (Pennington-Gray, Kerstetter and Warnick, 2002; Beldona, 2005; Beldona, Nusair and Demicco, 2009; Benckendorff, Moscardo and Pendergast, 2010; Li, Li and Hudson, 2013; Chiang *et al.*, 2014; Haydam *et al.*, 2017; Huang and Lu, 2017; Corbisiero and Ruspini, 2018; Monaco, 2018; Corbisiero, 2020). Sociologically, it is possible to argue that the pandemic events have strongly contributed to influencing the behavior of people as consumers and tourists, in particular of the younger people, who are experiencing a vision of the world which has been altered by COVID-19. In other words, their consumer lifestyle choices and vision of the world differ from those of previous generations since there are growing up in an era in which "make tourism safe" is gaining ground as new tourists paradigm. Many of the younger people have not yet had the opportunity to practice tourism on their own or have had limited travel experiences. Consequently, for them the tourism experience conditioned by COVID-19 will be something that they will naturally learn as a new normal.

As we have seen, the pandemic has generated economic and social problems, but it has also stimulated the formulation of alternative solutions both to fight the virus and to encourage a gradual recovery. These measures concerned two aspects in particular—safety and digitization—with strong impacts and consequences on the choices and behaviors of tourists.

As regards the first aspect, as has been explained in the various chapters of this book, with the arrival of the pandemic, safety has become one of the main concerns for people and institutions. Around the world, the attention to monitoring people's health has grown. At the same time, citizens and travelers have begun to demand greater health safety measures in all areas of life, including tourism. All the companies that work in the tourism sector have had to scrupulously follow the regulations of the WHO as well as the instructions and protocols set down by individual governments. Hotels, restaurants, travel companies, shops and travel agencies have suddenly found themselves needing to learn new ways of instilling a sense of security among tourists. The main hygienic measures to ensure people's safety have been hand sanitation, environmental disinfection, sanitation of all objects that are touched frequently and physical distancing. Customers and staff have had to learn to comply strictly with basic protective measures, and learn to use devices, such as protective masks or helmets. Thus, COVID-19 has started a new era, in which the tourist industry is obliged to guarantee health and safety to customers, through new protocols and new procedures. According to the French economist Attali (2020), all the structures empathically responding to the new needs of consumers, answering to their need for safety, will be rewarded. On an empirical level, the forced closures, together with the added costs of managing and maintaining the locations and the reduction of customer seats per square meter, have already inevitably led to a reduction in revenue for all the actors involved in the tourist chain.

Over time, this can only be compensated by an increase in prices for tourists themselves. This means that tourism in the traditional sense could become more exclusive since the practices that were so widespread to lower the costs of the holiday have lost much of their appeal because they are not considered completely safe.

The most obvious example of this is represented by the practices of sharing costs, which in recent years had spread considerably in the tourism field (e.g., Tuttle, 2014; Deloitte, 2015; Hamari, Sjöklint and Ukkonen, 2015; Smith, 2016; Ranzini *et al.*, 2017; Bernardi and Ruspini, 2018). In other words, because of the pandemic, the widespread model of sharing proposed by the sharing economy is one of those behaviors that could be most affected by the consequences of the pandemic. The sharing economy consists of exchanging, sharing, bartering, trading or borrowing goods and services on a large scale also thanks to the most modern communication technologies (Lathi and Selosmaa, 2013; Martin, Upham and Budd, 2015). It is evident that this will necessarily have to be rethought to adapt to the COVID-19 times. In recent years it was credited with putting tourists in direct contact with local populations. Travelers from all over the world were able to stay at local facilities to immerse themselves in the daily lives of the residents, enjoying their typical food and drink, taking part in their events (village festivals, traditions, and so on), challenging the strict rules of capitalism in a redistributive logic. The pandemic has questioned two of the fundamental assumptions on which this type of practice is based: meeting and sharing, which are two of the very situations that must be generally avoided to contain the infection. The increase in people's mistrust helps explain why recent studies on the subject (e.g., Batool *et al.*, 2020; Statista, 2021; Chen *et al.*, 2021; Gerwe, 2021; Hossain, 2021) have reported that the use of shared services is destined to decline in the short-to-medium term. In fact, sharing is an antithetical concept to "physical distancing," a phrase that the pandemic has introduced into common language and habits.

It seems plausible to ask: how many people will be willing to share accommodation or means of transport with other people they do not know? How many people will be willing to stay in close contact with other tourists from all over the world whose health or hygiene habits are unknown? COVID-19 has influenced the quality of interpersonal relationships, generating a general distrust among tourists. Paradoxically, people today are all conditioned by the idea that their fellow travelers can be carriers of diseases and infections. Even though the vaccine has been developed, the solution to eradicate mistrust toward strangers still does not exist. It will take a long time before resuming even intertraveler relationships completely. The new rules imposed by the pandemic have implicitly introduced at the social level a new cultural paradigm leading to a revisited normality in which new social and behavioral models are making their way. Under these circumstances, Airbnb, for example, has decided to change its business model, focusing on long-term rentals, closer to traditional ones, instead of short-term rentals, which until before the pandemic represented the company's core business. Similarly, today more and more restaurants are relying on a home delivery service of their food, knowing well that people and travelers more often than before prefer to eat at home instead of staying in an indoor confined public place (e.g., Aday and Aday, 2020; Hawkes, 2020; Hobbs, 2020; Gavilan *et al.*, 2021).

Being closely linked with these aspects, another element is emerging as a consequence of the new pandemic-produced scenario: a kind of homologation of the tourist offer, which necessarily guarantees certain standards to be taken into consideration by the majority of the world population. In the postmodern era, some social and cultural contexts, even if precarious or unsafe, were also considered interesting for their uniqueness. Many Asian countries were chosen by travelers for their typical open-air markets where animal meat was traded (not necessarily to be consumed). Vietnam, China, Korea are just some of the countries where the wildlife industry is a distinctive cultural trait. The fear of the spread of diseases and viruses, after the suspicion that the new coronavirus may have spread from Wuhan's wet market, has put a stop to these traditions. In fact, China announced a ban on consuming (and trading) of wild animals and Vietnam followed a ban on both the trade and consumption of wild animals. Many other traditions could disappear from different parts of the world. For example, we do know that a certain aversion toward bats and rats has intensified thanks to COVID-19. The destinations that are historically considered cities that host many of these animals (such as Bali, but also New York) should free themselves from this image if they intend to again attract the high number of tourists who are conditioned by a fear of these animals.

A similar argument can be made for the most polluted cities in the world. People have stopped taking them into consideration as tourist destinations for the fear that unhealthy air will negatively affect their health. Although the health risks associated with pollution have always been known and environmental causes have had a strong consensus around the world in the last few years, the shock of COVID-19 was necessary to influence the choices of many travelers globally, who now seem to be more worried and demanding because they have understood that their health can be endangered at any moment.

Thus, territories that still want to focus on tourism in general will have to work hard to eliminate or mitigate any elements that can be perceived as unsafe or dangerous by travelers. To encourage the development of these new perspectives, the creation of collaborative links among public institutions and the private sector will be of fundamental importance, thereby creating strong territorial synergies and developing a common ground for strategy planning. In other words, a definition of such a scenario must be the starting point for tour operators and for the governments programming the relaunch of the tourism sector, designing new possible entertainment solutions or improving existing tourism products, refining their own strategies of communication and promotion.

For example, the less crowded destinations, even though they are less known, can gain appeal in the eyes of tourists, because they allow people to stay outdoors with fewer people, reducing the opportunities for indoor meeting and gathering (e.g., Galvani, Lew and Perez, 2020; Spenceley, 2020; Higgins-Desbiolles, 2021). Slow tourism and domestic tourism seem to be two of the main tourist niches that may guarantee a gradual restart of physical mobility. Conceptually, slow tourism is promoted as a sustainable alternative to mass tourism. In fact, it offers a relaxed "zero kilometer" holiday, far from the hustle and bustle that usually characterizes travel experiences, respecting not only traditions and people, but also the environment and the ecosystem. In this scenario, villages, small towns and mountain or countryside locations certainly have an evident competitive edge.

Thus, the demand for nearby vacation locations, historically associated with lower sociodemographic status and older age brackets (Berrino, 2011), could probably be one of the new characteristics of future tourism, capable of undermining the mass tourism processes that had affected many cities during the prepandemic period.

We have already explained all the motivations making more and more people stay inside their own countries, and practice domestic tourism, which seems more reassuring (e.g., Das and Tiwari, 2020; Hussain and Fusté-Forné, 2021; Woyo, 2021). For this reason, it can be an important starting point for reactivating tourism in various territories. Each location should try to encourage its citizens, and the people who live in the neighboring areas, to practice forms of domestic tourism to discover the local beauties that are not too far from the travelers' homes, at least until international travel gradually returns to the tourist market, as happened before the pandemic. Each destination must provide information to travelers and build up relationships based on mutual trust, so they can be prepared for a gradual recovery.

To make this form of tourism truly competitive and sustainable, the wealth and beauty of the various territories need to be properly assessed, not only in terms of natural elements, such as sea, rivers or lakes, but also cultural features, such as local customs and traditions. On this basis, many other small tourist niches can be encouraged, such as sports tourism, food and wine tourism, heritage tourism and cultural tourism.

Putting aside the obvious and indisputable damage caused by the pandemic, its existence can perhaps also be interpreted as an opportunity to reset the past, offering the chance to start a new way of approaching the world of tourism, based specifically on the pursuit of sustainability. The new perspective must not exclusively concern tourist resorts and the offer of innovative and greener goods and services, but also, and above all, the development of a more ecological consciousness on the part of all travelers and citizens, who must be educated to be more attentive both to environmental aspects and to the enhancement of the culture and traditions of destinations and to direct their interest also toward locations other than big cities.

This does not mean that the great urban centers of the world have to be removed from the tourist circuit. The dynamism and abundance of cultural stimuli have represented and will continue to be the main elements underlying the charm of the most famous metropolises in the world (e.g., UNWTO, 2012; Bock, 2015; IPK International, 2015; Milano and Koens, 2021). However, sociologists, urban planners, demographers and tourism analysts have highlighted how these have been hit by growing tourism that has certainly brought economic benefits, but at a high price. The prices of rents and primary goods have increased, accompanied by a decrease in the quality of life, especially in all those cases in which the cities have gone far beyond their tourist-carrying capacity.

In Europe, in large capitals such as Berlin, Paris, Rome, Lisbon and Madrid, for example, fierce competition for short-term rental properties for tourists has made thousands of homes too expensive for middle-class local residents (e.g., O'Sullivan and Decker, 2007; Freytag and Bauder, 2018; Lestegás, Seixas and Lois-González, 2019; Ardura Urquiaga, Lorente-Riverola and Ruiz Sanchez, 2020; Yeager, Bynum Boley and Goetcheus, 2020). A similar situation has also been found in New York, where thousands of inhabitants have left their homes to return to their families in their countries of origin (e.g., Coles *et al.*, 2017). In Japan, the population increased in recent years only in Tokyo, due to internal migration from other parts of the country, while the rest of Japanese regions recorded a demographic decrease (Japanese Statistics Bureau, 2020). Under these circumstances, the government established a regional revitalization task force to focus on regional hub cities' development (e.g., Feng, 2015).

Actually in the general reorganization of the tourism sector, an opportunity to imagine the city as a tourist destination is in place, but it must balance the needs of the tourism market with the life and needs of local citizens (Miller, 2021).

As I predicted, another element that has characterized the pandemic period has been digitization. Globally, an acceleration in the technological world has taken place. Many technological schemes have been created to allow people from all over the world to know in real time where, and under which conditions they can travel. Similarly, following border closures, many opportunities have been created to enable tourists to start having remote connections with destinations, while they wait to visit these places physically as soon as possible. In this sense, territorial and tourism marketing has taken on a very central role. Within an increasingly globalized tourism market, all the territories of the world have begun to identify the most suitable strategies to stimulate and attract future travelers, through the sharing of interactive contents such as images and videos, virtual guided tours, quizzes on typical products or on territorially identifying objects. This digitization has produced two effects: travelers who really intend to reach a destination in the postpandemic era are able to get a taste of the experience by first identifying areas and activities of interest to them; people who do not intend to visit that destination in person are still able to practice stationary tourism, traveling without moving through physical space. In both cases, the proposed contents were conveyed online, allowing destinations to increase their reputation globally, regardless of the immediate exploration of the destination or not.

Furthermore, many other tools have been developed specifically to increase safety levels, to ensure better sanitation and to identify antipandemic strategies in a smart way. This type of approach was taken early on in the pandemic crisis by many countries, lowering the contagion curve in record time. We have witnessed a proliferation of apps, platforms and internet sites that have exploited artificial intelligence to achieve this important milestone. A technology at the service of the public and safeguarding their health has been developed. This includes not only online services and communication tools, but also wearable devices that allow doctors and patients to interact remotely. These are biosensors inserted on clothing, watches (smartwatch), T-shirts, shoes, trousers, belts, bands (smart clothing), glasses (smart glasses) capable of detecting and measuring various biological parameters (heart rate, respiratory rate, saturation oxygen, body temperature, blood pressure, glucose, sweat, breath and brain waves) and providing information on the state of health and even on the lifestyle of an individual (physical activity, sleep, nutrition and calories consumed).

Indeed, an Indian Institute of Science research team has implemented the COVID-19 Sounds app to create algorithms that allow people to recognize automatically whether a cough is caused by COVID-19, distinguishing it from one caused by other ailments (e.g., Sharma *et al.*, 2020). Using a computer model based on a neural network, scientists were able to detect the disease by analyzing the participants' cough and breath recordings. This is just one of many examples that have projected the world into a more technological future.

With particular regards to the tourism sector, technologies have not only helped travelers navigate the new geography of the world, but have been a great aid in automating controls and keeping off health threats. From thermal scanners to smart cameras, several innovations have been formulated.

The wide application of antipandemic technologies and digitization in the contemporary world will have to represent a starting point but not an arrival point, motivated by the wish to continue investing in research and technological innovation on a global level. In other words, research will have to be increasingly understood as an investment but not an expense that is a burden on public funding in order to have the possibility of building a better world by learning from the mistakes of the past.

Obviously, if everyone in the world is to benefit from tourism, it is essential to do something about the digital divide, which is an obstacle to social and individual growth. From this point of view, the pandemic has made it even more evident that the quantity and quality of access to the digital world has caused serious inequality, which creates a gap between the countries and people who can rely on technologies and infrastructures and those who can't.

In this regard, one of the necessary future challenges on the part of institutions around the world will be to take on the task of encouraging digitization and digital literacy, to foster the integration of tradition and novelty, intervening, in particular, on the different levels of the "digital divide," which are today present globally (e.g., van Dijk, 2005; Hargittai, 2010; Helsper, 2012). As underlined in the critical literature dealing with media education, not all subjects have the concrete opportunity to use new technologies or have high-performance connections (e.g., Castells, 2001; Norris, 2001).

Furthermore, it is not certain that those who do own devices and can surf the net then have the necessary skills to use new technologies adequately and consciously (e.g., Hargittai, 2002; Zillien and Hargittai, 2009; van Deursen and Helsper, 2015).

On these issues it is important that institutions devise new policies to familiarize all classes of the population with new technologies, in general, and, more specifically in the field of tourism. This type of commitment is necessary to impact on social inclusion and counter the new forms of marginalization that the pandemic has accentuated.

Still on the level of citizens' rights and the guarantee of these for everyone globally, a final consideration to be made concerns vaccines and the so-called immunity passports. As is known to all, international vaccine certificates are not a novelty. For more than 80 years, there have in fact been, in various forms, documents containing health information that travelers must carry with them under certain circumstances. The most famous and recognized is the International Certificate of Vaccination and Prophylaxis, known as the "Yellow card," which is a document written in several languages and issued by the World Health Organization or by the national health authorities to certify vaccination against specific diseases. This card is internationally recognized and is mandatory for entering some countries where there are higher risks posed by the spread of specific diseases. The new so-called Green Pass is not very different from the Yellow card, but it aims to simplify its management with the aid of apps and online platforms. The discussion about the legitimacy of the "immunity passport" has involved an increasing number of governments, health institutions, but also some large companies, especially in the digital technology sector. In particular, the countries whose economies depend heavily on tourism have supported the scheme, considering the passport of immunity as a possible tool for relaunching the industry. At the same time the idea of an immunity passport has aroused controversy from the start for both its practicality and the risks associated with the protection of privacy of privileged countries. Furthermore, opponents have argued that the effectiveness of the vaccination campaign especially on variants of the virus is still poorly measurable, so proof that vaccines exert a containment effect over the coronavirus are not enough. Finally, there are some difficulties in setting universal standards.

As just mentioned, the immunity passport has aroused some perplexity among observers and experts in digitalization and privacy. Being the most sensitive, health data must be processed with special care, especially if transmitted and exchanged online. Many detractors of the immunity passport don't uphold the idea that these data could be managed through proprietary applications. Also, for this reason, the Linux Foundation, the foundation that supports the development of the Linux open-source computer system, proposed the development of an open solution with a freely accessible code in order to offer more verification systems and ensure greater transparency. The main orientation has been to create platforms and applications with the function of managing immunity passports, making it easier and faster to check people's certificates at the borders and especially at airports before boarding.

In this scenario, the IATA developed an "IATA Travel Pass," a mobile app to help travelers easily and safely manage their trips in line with government requirements for COVID-19 testing. On the traveler's mobile device, the app stores encrypted data, including the results of tests or the records of vaccinations. Travelers can choose which information to share from their phones with the airlines and the local authorities of the destinations they intend to visit. The IATA Travel Pass also follows the highest standards of data protection laws, including the European Union General Data Protection Regulation, guaranteeing maximum levels of data confidentiality. This has been developed in line with the principles of Self-Sovereign Identity (SSI).

To test the pass, the IATA started a partnership with Singapore Airlines, which, since December 23, 2020, offered the Travel Pass option on flights to the city-state departing from Kuala Lumpur and Jakarta, with a list of clinics selected in the capitals of Indonesia and Malaysia where it was possible receive a test for COVID-19 and obtain the digital health certificate to travel. In the early months of 2021, Etihad, Air New Zealand, Qatar Airways, Emirates and Malaysia Airlines also started experimenting with the innovation. Thanks to a collaboration with the government of Panama, Copa Airlines was the first airline outside the Asia-Pacific region to test the service, followed by the first African company, RwandAir.

Globally, the presence of various other proposals for the management of immunity passports has certainly made the situation rather chaotic, complicating the identification of a single solution shared by most of the subjects involved, both at the institutional and private levels.

For example, in Asia, China launched the world's first vaccine passport to allow Chinese citizens to travel in early March 2021. This was a nonmandatory digital certificate that could be accessed through the WeChat social platform, and it was launched by the Department of Consular Affairs of the Foreign Ministry. The Chinese Vaccination Passport includes information on nucleic acid tests and serum antibody results, vaccinations and other related information. Foreign Minister Wang Yi said that in the near future, as more and more countries agree on the need for mutual recognition of health certificates with China, this international travel document will play a greater role in promoting the healthy and safe movement of people and provide Chinese citizens with a solid guarantee when traveling abroad. According to the state agency Xinhua, the Chinese program includes an encrypted QR code that allows each country to obtain information on the health of travelers. As we have previously explained, this technology is already present and widespread within WeChat and other Chinese smartphone apps to allow people access to national transport and many public spaces.

In early 2021, the idea of a "green" certification appealed to the new American administration. In the United States, the recommendations from the Centers for Disease Control and Prevention (CDC) indicated that people who had completed the vaccination process could meet vaccinated friends and relatives indoors without the need for masks or social distancing and even meet unvaccinated people belonging to a single family without having to worry about anti-COVID-19 prevention measures.

Similarly, the European Commission has worked to implement a Digital Green Certificate to allow all the European Union (EU) member countries a centralized tool through which to monitor the state of people's health, travelers in particular. From July 2021 the EU Green pass has been recognized in all 27 countries of the European Union, and open to Iceland, Liechtenstein, Norway and Switzerland. It has been issued to EU citizens and their families regardless of their nationality, as well as to citizens of third countries residing in the EU and to visitors who have the right to travel to other member states. This measure has been defined as a temporary solution that will be suspended once the World Health Organization declares the end of the international health emergency. The EU Digital Green Certificate considers three possible parameters to give people the opportunity to travel: vaccination, negativity to a test (NAAT/RT-PCR test or rapid antigen test) or recovery from COVID. Thus, it has been implemented to allow travel not only for people who have been recognized as immune, but also for people who have tested negative to the virus and immune to the disease. As for vaccines, the people who received a shot from pharmaceutical companies authorized by the EU could certainly have the Digital Green Certificate. The pass has been designed to be universally effective among EU countries, ensuring that the different types of digital green certificates (vaccination status, test results, cure status) are standardized according to mutually agreed policies, rules and specifications. In other words, this means that a certificate issued in one member state has to be recognized in another. Each certificate is made up of to include a limited set of information such as name, date of birth, date of issue, relevant vaccines, treatment information about its owner, and a unique certificate identifier. In terms of privacy, during a possible verification, the information to be checked could be only the validity and authenticity of the certificate. All health data remains in the member states that issued the certificate. European states have been placed in the conditions of freely decide if the certification could be considered a mandatory pass to access to public places, including leisure facilities (theaters, cinemas and stadiums), and work and study places. In this context, some European countries have imposed the obligation of the green certificate to carry out some of these activities or for people who work in contact with the public, as well as for healthcare personnel. The first country that took this decision has been France.

Similarly, in early 2021, Israel and Denmark had already launched a sort of vaccination passport to certify immunization or recovery from the disease. This kind of solution allowed its owners to return to enjoy a few of the privileges of prepandemic life. Thus, restaurants, bars, gyms, swimming pools and many other services had become accessible only to green pass holders.

Likewise, in March 2021 in the United Kingdom, the shipping company P&O Cruises made it known that it would have resume domestic cruise holidays in summer 2021, but limiting it only to fully vaccinated against COVID-19 residents. In other words, only English vaccinated people could take part in the short cruises organized between the end of June and September. The cruise ships Britannia and Iona received and welcomed passengers on trips of 3–7 nights around the United Kingdom, departing from Southampton. Due to the COVID-19 restrictions ships not called at any port, although there was the usual on-board dining and entertainment program. The cruise line, owned by the Carnival group, said that all passengers had to receive all doses of the vaccine for full immunization at least 7 days prior to departure in order to be allowed to embark. Those who failed to provide evidence not were able to board the ship, while the crew was subjected to quarantine and received regular tests on board in advance.

These solutions, which in fact limit the possibilities for everyone who has not yet received their dose of the vaccine, lead to a series of further considerations.

As we know, the vaccination campaign in the world has been characterized from the beginning by some obstacles and not a few delays. We have witnessed a vaccine rush, influenced above all by the fact that the population has been divided into different health risk categories, giving priority to the elderly, the sick, doctors, law enforcement officers and personnel in activities that are considered essential or of public and social utility. So, even though the entire world population must be guaranteed the vaccine, we have a situation in which there are people who tend to be immune and others who wait to receive their injection. The vaccinated population is different in each country. This phenomenon is generating a further disparity among people, discriminating against people who can gradually resume living a regular life and those who are still forced to comply with the strict rules imposed by the virus.

Furthermore, we have long explored how the boundary between public interest and individual freedom is very critical and problematic.

On this regard, we must specify that the hierarchy of the population on the basis of the vaccine certainly represents an unprecedented scenario in terms of individual rights. It is important to monitor the state of health of the world population, and it is also necessary to identify the most suitable strategies to avoid penalizing the people who obtain the vaccine at a later stage. Consequently, making the Green Pass necessary to take part to some social activities can be considered a temporary palliative to restore a global normality, linked to the health emergency and the complete implementation of the vaccination plan, but not a definitive solution.

Moreover, at the moment we lack certain data on the validity of the immunity of the vaccine or about people recovering fully from COVID-19. Indeed, it is uncertain whether protection is permanent or not. In empirical reality this means that any immunity license will inevitably be destined to expire. For example, Israel has been the first country to consider the vaccine valid for 6 months, administering the third dose of Pfizer vaccine on the elderly and fragile population starting from August 2021. Prime Minister Naftali Bennett underlined that this decision has been made by a team of experts after careful analysis and research. Consequently, linking the possibility of traveling and accessing public services and places to the temporal validity of the vaccine would mean giving an expiry date to individual freedoms, to be renewed periodically.

Finally, an unforeseen consequence of this new, chaotic and complex situation is what we can define as "vaccine tourism," an emerging phenomenon that affects some people around the world who travel to other countries to get a dose of the vaccine.

As reported by The Guardian (Misra, 2021), in the United States, for example, there have been over 50 unique vaccination plans. In Wisconsin, mink farmers have been considered among the people with the highest priority to receive the vaccine. In New Jersey, smokers could have priority access to the vaccine. In Colorado, reporters fell into the category of frontline workers. This complex system has given rise to a new type of pandemic travel in which people started to cross state or even country borders to gain early access to vaccines. Due to the absence of a standardized protocol and the complexity of the American health system, thousands of people have obtained vaccines outside their home states. For example, in the first months of the vaccination campaign, 37,000 Americans residing elsewhere were vaccinated in Florida, arousing protests from local residents, so much so that Florida subsequently had to specify that it would start vaccinating only seasonal residents and part-time residents.

The Daily Telegraph (Carr, 2021) brought to light the rather sensational (but legal) case of Knightsbridge Circle, a luxury travel service in London that organized vaccine vacations for over 65s in the United Arab Emirates.

Similarly, there are already some tour packages online that promise to combine the travel experience with a shot of the vaccine. Currently, the cheapest one is from an Indian travel agency that offers a round trip to London, with a hotel stay, three excursions to explore the city and two shots of Pfizer at a cost of €1,700.

This situation has already stirred a lot of discussion, so much so that some governments have had to intervene to control the situation. The Dubai Tourist Office, for example, issued a statement to remind people that the rumors of vaccine tourism were groundless and that only people with a resident card could be vaccinated. The Government of Cuba had instead taken a strikingly opposite position, launching in 2021 a real promotional campaign for the vaccine called "Beach, Caribbean, Mojito and Vaccine." Cuban health authorities planned to produce 100 million shots of Soberana 2, to guarantee vaccination for all its population, but also for tourists.

So, tourism with vaccination as "the bait" is becoming a potential problem. If a vaccine-related tourism market were to develop in the next months, the risk that it would run is that immunization, and the rights it brings with it, are likely to become a privilege reserved to the few people who will be able to afford it. In the future, if the vaccine turns into a luxury item, then the opportunities for speculation are bound to multiply.

Bibliography

Aday, S and Aday, M S (2020) 'Impact of COVID-19 on the food supply chain', *Food Quality and Safety*, 4, 4: 167–180.

Alwin, D and McCammon, R (2007) 'Rethinking generations', *Research in Human Development*, 4, 3–4: 219–237.

Ardura Urquiaga, A, Lorente-Riverola, I and Ruiz Sanchez, J (2020) 'Platform-mediated short-term rentals and gentrification in Madrid', *Urban Studies*, 57, 15: 3095–3115.

Attali, J (2020) 'Que naîtra-t-il?', *Linkedin*. Retrieved from: https://www.linkedin.com/pulse/que-na%25C3%25AEtra-t-il-jacques-attali/?trackingId=cYufUofxQhWWp9DwZbK6YQ%3D%3D (accessed 04/04/2021).

Attias-Donfut, C (1991) *Générations et âges de la vie*, Paris: PUF.

Batool, M, Ghulam, H, Azmat Hayat, M, Naeem, M Z, Ejaz, A, Imran, Z A, Spulbar, C, Birau, R and Gorun, T H (2020) 'How COVID-19 has shaken the sharing economy? An analysis using Google trends data', Economic Research-Ekonomska Istraživanja, 34, 1: 2374–2386.

Beldona, S (2005) 'Cohort analysis of online travel information search behaviour: 1995–2000', *Journal of Travel Research*, 44, 2: 135–142.

Beldona, S, Nusair, K and Demicco, F (2009) 'Online travel purchase behaviour of generational cohorts: A longitudinal study', *Journal of Hospitality Marketing and Management*, 18, 4: 406–420.

Benckendorff, P, Moscardo, G and Pendergast, D (eds.) (2010) *Tourism and Generation Y*, Oxfordshire: Cabi International.

Bernardi, M and Ruspini, E (2018) 'Sharing tourism economy among millennials in South Korea', in Y Wang, A Shakeela, A Kwek and C Khoo-Lattimore (eds.) *Managing Asian Destinations*, Singapore: Springer.

Berrino, A (2011) *Storia del turismo in Italia*, Bologna: Il Mulino.

Bock, K (2015) 'The changing nature of city tourism and its possible implications for the future of cities', *European Journal of Futures Research*, 3, 20.

Carr, G (2021) 'Wealthy members of £25,000-a-year Knightsbridge-based travel concierge service who are aged over 65 are being flown to the United Arab Emirates and India on private jets to receive the Covid jab', *The Daily Telegraph*, 01/14/2021. Retrieved from: https://www.dailymail.co.uk/news/article-9147557/Travel-concierge-service-charging-members-40-000-private-Covid-jab.html (accessed 04/04/2021).

Castells, M (2001) *Internet Galaxy*, Oxford: Oxford University Press.

Chen, G, Cheng, M, Edwards, D and Xu, L (2021) 'COVID-19 pandemic exposes the vulnerability of the sharing economy: A novel accounting framework', *Journal of Sustainable Tourism*, 1: 1–18.

Chiang, L, Manthiou, A, Tang, L, Shin, J and Morrison, A (2014) 'A comparative study of generational preferences for trip-planning resources: A case study of international tourists to Shanghai', *Journal of Quality Assurance in Hospitality and Tourism*, 15, 1: 78–99.

Coles, P A, Egesdal, M, Ellen, I G, Li, X and Sundararajan, A (2017) 'Airbnb usage across New York City neighborhoods: Geographic patterns and regulatory implications', SSRN Electronic Journal, 10: 1–18.

Corbisiero, F (2020) 'Sostenere il turismo: Come il COVID-19 influenzerà il viaggio del future', *Fuori Luogo. Rivista di Sociologia del Territorio, Turismo, Tecnologia*, 7, 1: 69–79.

Corbisiero, F and Ruspini, E (eds.) (2018) 'Millennials and generation z: Challenges and future perspectives for international tourism', *The Journal of Tourism Futures-ETFI*, Special Issue, 4, 1.

Das, S S and Tiwari, D A (2020) 'Understanding international and domestic travel intention of Indian travellers during COVID-19 using a Bayesian approach', *Tourism Recreation Research*, 1: 1–17.

Deloitte (2015) *Collaboration Generation*, London: Deloitte.

ET Bureau (2020) 'Covidivorces, quaranteens and coronials: Why coronavirus puns and neologisms are going viral', *The Economist Times*, 04/09/2020. Retrieved from: https://economictimes.indiatimes.com/magazines/panache/covidivorces-quaranteens-and-coronials-why-coronavirus-puns-and-neologisms-are-going-viral/articleshow/75057681.cms?utm_source=contentofinterest&utm_medium=text&utm_campaign=cppst (accessed 04/06/2021).

Feng, L (2015) 'Spatial structure of aging in four regional central cities in Japan: Sapporo, Sendai, Hiroshima and Fukuoka', *Geography*, 61, 1: 43–62.

Freytag, T and Bauder, M (2018) 'Bottom-up touristification and urban transformations in Paris', *Tourism Geographies*, 20, 3: 443–460.

Galvani, A, Lew, A A and Perez, M S (2020) 'COVID-19 is expanding global consciousness and the sustainability of travel and tourism', *Tourism Geographies*, 22, 3: 567–576.

Gavilan, D, Balderas-Cejudo, A, Fernández-Lores, S and Martinez-Navarro, G (2021) 'Innovation in online food delivery: Learnings from COVID-19', *International Journal of Gastronomy and Food Science*, 1.

Gerwe, O (2021) 'The covid-19 pandemic and the accommodation sharing sector: Effects and prospects for recovery', *Technological Forecasting and Social Change*, 167: 120733.

Hamari, J, Sjöklint, M and Ukkonen, A (2015) 'The sharing economy: Why people participate in collaborative consumption', *Journal of the Association for Information Science and Technology*, 1: 1–28.
Hargittai, E (2002) 'Second-level digital divide: Differences in people's online skills', *First Monday*, 7, 4: 84.
Hargittai, E (2010) 'Digital na(t)ives? Variation in Internet skills and uses among members of the net generation', Sociological Inquiry, 80, 1, 93–113.
Harmony, M. (2020) 'Coronnials', *Urban Dictionary*. Retrieved from: https://www.urbandictionary.com/define.php?term=coronnials (accessed 04/04/2021).
Hawkes, C (2020) 'COVID-19 and the promise of food system innovation', in J Swinnen and J McDermott (eds.) *COVID-19 and Global Food Security*, Washington: International Food Policy Research Institute.
Haydam, N, Purcărea, T, Edu, T and Negricea, C (2017) 'Explaining satisfaction at a foreign tourism destination—An intra-generational approach evidence within generation y from South Africa and Romania', *Amfiteatru Economic*, 19, 45: 528–542.
Helsper, E J (2012) 'A corresponding fields model for the links between social and digital exclusion', *Communication Theory*, 22, 4: 403–426.
Higgins-Desbiolles, F (2021) 'The war over tourism: Challenges to sustainable tourism in the tourism academy after COVID-19', *Journal of Sustainable Tourism*, 29, 4: 551–569.
Hobbs, J E (2020) Food supply chains during the COVID-19 pandemic, *Canadian Journal of Agricultural Economics*, 68: 171–176.
Holbrook, M B and Schindler, R M (1989) 'Some exploratory findings on the development of musical tastes', *Journal of Consumer Research*, 16, 1: 119–124.
Hossain, M (2021) 'The effect of the COVID-19 on sharing economy activities', *Journal of Cleaner Production*, 280, 1: 124782.
Huang, W and Lu, Y (2017) 'Generational perspective on consumer behaviour: China's potential outbound tourist market', *Tourism Management Perspectives*, 24: 7–15.
Hussain, A and Fusté-Forné, F (2021) 'Post-pandemic recovery: A case of domestic tourism in Akaroa (South Island, New Zealand)', *World*, 2, 1: 127–138.
IPK International (2015) *Tourismus als globaler Wachstumstreiber*, Berlin: IPK International.
Japanese Statistics Bureau (2020) *Japan Statistical Yearbook: Population and Households*, Tokyo: Ministry of Internal Affairs and Communication.
Ladwein, R, Carton, A and Sevin, E (2009) 'Le capital transgénérationnel: Une transmission dynamique des pratiques de consommation entre deux générations dans un cadre familial', *Recherche et Applications en Marketing*, 24, 2: 1–27.
Lathi, M and Selosmaa, J (2013) *A Fair Share: Towards a New Collaborative Economy*, Helsinki: Atena.
Lestegás, I, Seixas, J and Lois-González, R C (2019) 'Commodifying Lisbon: A study on the spatial concentration of short-term rentals', *Social Sciences*, 8, 2: 33.
Li, X, Li, X R and Hudson, R (2013) 'The application of generational theory to tourism consumer behaviour: An American perspective', *Tourism Management*, 37: 147–164.
Martin, C J, Upham, P and Budd, L (2015) 'Commercial orientation in grassroots social innovation: Insights from the sharing economy', *Ecological Economics*, 118: 240–251.

McLaren, D and Agyeman, J (2015) *Sharing Cities: A Case for Truly Smart and Sustainable Cities*, Cambridge: MIT Press.

Meyer, G and Shaheen, S (eds.) (2017) *Disrupting Mobility, Impacts of Sharing Economy and Innovative Transportation on Cities*, London: Springer.

Milano, C and Koens, K (2021) 'The paradox of tourism extremes. Excesses and restraints in times of COVID-19', Current Issues in Tourism, 1: 1–13.

Miller, D S (2021) 'Abrupt new realities amid the disaster landscape as one crisis gives way to crises', *Worldwide Hospitality and Tourism Themes*, 13, 3: 304–311.

Misra, K (2021) "Vaccine tourism': tens of thousands of Americans cross state lines for injections', *The Guardian*, 01/30/2021. Retrieved from: https://www.theguardian.com/us-news/2021/jan/31/us-vaccine-tourism-state-borders-covid-19-shots (accessed 04/04/2021).

Monaco, S (2018) 'Tourism and the new generations: Emerging trends and social implications in Italy', *Journal of Tourism Futures*, 4, 1: 7–15.

Norris, P (2001) *Digital Divide: Civic Engagement, Information Poverty and the Internet in Democratic Societies*, Cambridge: Cambridge University Press.

O'Sullivan, E and Decker, P (2007) 'Regulating the private rental housing market in Europe', *European Journal of Homelessness*, 1: 95–117.

Pennington-Gray, L, Kerstetter, D L and Warnick, R (2002) 'Forecasting travel patterns using Palmore's cohort analysis', *Journal of Travel and Tourism Marketing*, 13, 1–2: 125–143.

Ramkissoon, H (2020) 'Perceived social impacts of tourism and quality-of-life: A new conceptual model', *Journal of Sustainable Tourism*, 1: 1–17.

Ranzini, G, Newlands, G, Anselmi, G, Andreotti, A, Eichhorn, T, Etter, M, Hoffmann, C, JJrss, S and Lutz, C (2017) 'Millennials and the sharing economy: European perspectives', SSRN Electronic Journal: 1–26.

Sharma, N, Krishnan, P, Kumar, R, Ramoji, S, Chetupalli, S R, Ghosh, P K and Ganapathy, S (2020) 'Coswara. A database of breathing, cough, and voice sounds for COVID-19 diagnosis', *Proceeding of Interspeech*, 27, 60: 4811–4815.

Sheller, M (2018) *Mobility Justice: The Politics of Movement in an Age of Extremes*, London: Verso.

Smith, A (2016) *Shared, Collaborative and On Demand: The New Digital Economy*, Washington, D.C.: Pew Research Center.

Spenceley, A (2020) *Building Nature-Based Tourism Back from COVID-19: Recovery, Resilience and Sustainability*, Gland: Luc Hoffmann Institute.

Statista (2021) 'How will your use of sharing economy services change after the containment of Covid-19?'. Retrieved from: https://www.statista.com/statistics/1110632/sharing-economy-services-united-states-covid19/ (accessed 04/04/2021).

Tuttle, B (2014) 'Can we stop pretending the sharing economy is all about sharing?', *Money* 1: 14–30.

UNWTO (2012) *Global Report on City Tourism*, Madrid: UNWTO.

van Deursen, A J and Helsper, E J (2015) 'The third-level digital divide: Who benefits most from being online?', *Communication and Information Technologies Annual*, 10: 29–53.

van Dijk, J (2005) *The Deepening Divide: Inequality in the Information Society*, London: Sage.

Ward, S (1974) 'Consumer socialization', *Journal of Consumer Research*, 1, 2: 1–17.

Woyo, E (2021) 'The sustainability of using domestic tourism as a post-COVID-19 recovery strategy in a distressed destination', *Information and Communication Technologies in Tourism*, 1: 476–489.

Yeager, E, Bynum Boley, B and Goetcheus, C (2020) Conceptualizing peer-to-peer accommodations as disruptions in the urban tourism system, *Journal of Sustainable Tourism*, 12: 1–16.

Zillien, N and Hargittai, E (2009) 'Digital distinction: Status-specific types of internet usage', *Social Science Quarterly*, 90, 2: 274–291.

Zwanka, R J and Buff, C (2021) 'COVID-19 generation: A conceptual framework of the consumer behavioral shifts to be caused by the COVID-19 pandemic', *Journal of International Consumer Marketing*, 33, 1: 58–67.

Index

Note: *Italicized* pages refer to figures. Page numbers followed by "n" refer to notes.